Praise for *Our Religious Brains: What Cognitive Science Reveals about Belief, Morality, Community and Our Relationship with God*

"A fresh look at ancient spiritual questions through the informed lens of modern neuroscience. A fine contribution to the dialogue between science and religion."

—**Dr. Daniel S. Levine**, professor of psychology, University of Texas at Arlington

"Invites laity and clergy into a fascinating and dynamic conversation at the intersections of neuroscience and religion. Helps us revise and expand our understanding of once familiar ideas about the soul, spiritual experience, the frailty of our morality, prayer and the longing for community so that we return to our religious lives with deepened appreciation."

—**Nancy Ramsay**, executive vice president, dean, and professor of pastoral theology and pastoral care, Brite Divinity School

"Easy to read, very wise meditations ... that manage to respect both science *and* religion. Rabbi Mecklenburger deftly guides us through current research, ancient teaching and ultimately assembles the elements of an evolved twenty-first-century Jewish theology."

—**Lawrence Kushner**, Emanu-El Scholar, Congregation Emanu-El of San Francisco; author, *I'm God; You're Not: Observations on Organized Religion and Other Disguises of the Ego*

"An invaluable and immensely relevant book. It will be most useful to people of all religious communities."

—**Kenneth Cracknell**, former Michael C. Gutteridge Chair of Systematic and Pastoral Theology, Wesley House, Cambridge; retired professor of Theology and Mission, Brite Divinity School

"Sets the foundations for a rethinking of the relationship between the astonishing findings of brain science and the ways in which religious people believe and behave. Gives readers with both spiritual and scientific inclinations a wealth of new directions to explore."

—**Susan White**, Alberta H. and Harold L Lunger Professor Emerita of Spiritual Resources, Brite Divinity School

"A profound reflection on the very essence of who we are as human beings. A genuinely original and thought-provoking work, and religious adherents of every stripe will have their faith deepened and broadened through a reading of these pages. I recommend it enthusiastically!"

—**Dr. David Ellenson**, president, Hebrew Union College–Jewish Institute of Religion

"The very first to tackle the complex and fascinating issues of the relationship between brain science and Judaism. The issues raise[d] are on the intellectual and spiritual frontier that all Jewish thinkers will begin to explore in the coming years. [These] pioneering questions leave the reader with an abundance of food for thought!"

—**Rabbi David Nelson**, campus rabbi and visiting assistant professor of religion, Bard College; author, *Judaism, Physics and God: Searching for Sacred Metaphors in a Post-Einstein World*

"A significant contribution to the communal conversation on the nature of reality, human wiring and the religious impulse."

—**Rabbi Elie Kaplan Spitz**, author, *Does the Soul Survive? A Jewish Journey to Belief in Afterlife, Past Lives & Living with Purpose*

"A well-written and highly accessible introduction to an important subject. New research on human consciousness has much to contribute to our understanding of ourselves and our striving toward the mystery we call God."

—**Dr. Arthur Green**, Irving Brudnick Professor of Philosophy and Religion, Hebrew College; author, *Ehyeh: A Kabbalah for Tomorrow*

"A serious exploration of the religious experience, drawing on classic theological sources as well as modern-day findings of cognitive science.... This is an interesting book that will make readers think—not a small accomplishment!"

—**Rabbi Marc D. Angel**, founder and director, Institute for Jewish Ideas and Ideals; author, *Maimonides, Spinoza and Us: Toward an Intellectually Vibrant Judaism*

Our
Religious
Brains

Our Religious Brains

What Cognitive Science Reveals about Belief, Morality, Community and Our Relationship with God

RALPH D. MECKLENBURGER

FOREWORD by Dr. Howard Kelfer
PREFACE by Dr. Neil Gillman

For People of All Faiths, All Backgrounds

JEWISH LIGHTS Publishing

Woodstock, Vermont

Walking Together, Finding the Way®

SKYLIGHT PATHS®
PUBLISHING

Woodstock, Vermont

Our Religious Brains:
What Cognitive Science Reveals about Belief, Morality, Community and
Our Relationship with God

2012 Hardcover Edition, First Printing
© 2012 by Ralph D. Mecklenburger
Foreword © 2012 by Howard Kelfer
Preface © 2012 by Neil Gillman

Library of Congress Cataloging-in-Publication Data
Mecklenburger, Ralph D.
Our religious brains : what cognitive science reveals about belief, morality,
community and our relationship with God / By Ralph D. Mecklenburger ;
foreword by Neil Gillman.
p. cm.
Includes bibliographical references and index.
ISBN 978-1-58023-508-2
1. Brain—Religious aspects. 2. Psychology, Religious. 3. God. 4. Cognitive science. 5. Religion and science. I. Title.
BL53.M44 2012
201'.615—dc23
2011052402

10   9   8   7   6   5   4   3   2   1

Manufactured in the United States of America
Jacket and interior design: Heather Pelham

Published by Jewish Lights Publishing and SkyLight Paths Publishing
Divisions of Longhill Partners, Inc.
Sunset Farm Offices, Route 4, P.O. Box 237
Woodstock, VT 05091
Tel: (802) 457-4000   Fax: (802) 457-4004
www.jewishlights.com   www.skylightpaths.com

For Ann, Elissa, and Alan

Contents

Foreword

The brain-mind dichotomy. The brain consciousness dilemma. The brain-soul mystery. For how long has humankind pondered these apparent unknowables?

Is the mind embodied in a little homunculus in the pituitary gland? If I really think, does that automatically mean I am? If I don't think much anymore—I have software and thumb drives for that—does that mean I am not?

But surely we are getting closer. We know we each are wired with about one hundred billion neurons. There must be some threshold number of linkages in that wiring diagram from which consciousness arises.

And if consciousness arises from interconnected brain cells, surely so do our personalities, our perceptions of our world, and our responses to it.

Isn't it only logical, therefore, that we can at least ask the question: from where does the spirit, the soul—even God—arise?

In *Our Religious Brains*, Rabbi Ralph Mecklenburger asks these questions. In working with him, I have come to understand that he genuinely feels the need, from the perspective of a practicing liberal Jewish theologian, to make a timely and well-thought-out argument: our evolving knowledge of the functions of our conscious and unconscious brain have immense bearing on our understanding of the most basic religious tenets, up to and including our understanding of God.

Rabbi Mecklenburger makes no pretenses that he is anything beyond a practicing rabbi—from my perspective, a well-read, thoughtful, and very intelligent one—who has the insight to connect meaningfully with what his congregants and what open-minded people in general are thinking about. He does not try to emulate a junior neuroscientist; he does not remotely try to connect God to dopamine receptors, calcium channels, or the latest fad in functional MRI.

One could even argue that modern religious thinkers have an obligation to update—or at least review—their beliefs in light of what we are learning about brain science. Who is in the best position to do this but a practicing theologian? *Our Religious Brains* could never have been written as an undergraduate project or even as a doctoral thesis. It is only through a lifetime immersed in studies of theology and cognitive neuroscience on the one hand, and in the daily travails of congregants on the other, that such a work could have evolved.

Finally, it seems that scientists, particularly brain scientists, are publishing by the dozens their own treatises about how to connect their brand of science with religion, philosophy, metaphysics, and more. Is it not hypocritical to think that a theologian cannot bring a little science into his own work? Rabbi Mecklenburger's arguments are timely, well researched and well considered. They do not require an imaginative leap that may be based on speculation or any misinterpretation of the limited amount of neuroscience that he brings to bear on his thesis.

In summary, *Our Religious Brains* is a gem. Read it for the information it contains. Better yet, digest the words and thoughts, deliver them to your conscious brain, and make your own connections.

—Dr. Howard Kelfer,
Department of Neurology,
Cook Children's Hospital, Fort Worth, Texas

Preface

My first encounter with the issues raised by neuroscience stemmed from my discussion of resurrection and immortality in my book *The Death of Death: Resurrection and Immortality in Jewish Thought* (Jewish Lights). I was trying, then, to understand the classical Jewish teaching that our souls never die, that they enjoy immortality simply because of the nature of souls. This led me to ask questions such as, "So what is a soul?" And then, "What is the relation between soul and mind?" Or "soul, mind, and brain?" "What is the source of my sense of self?" And finally, "What is consciousness?" I recall leaping out of my chair when American neuropsychiatrist and Nobel Prize winner Eric Kandell responded to a question posed by acclaimed broadcast journalist Charlie Rose, "Mind is what the brain does."

So I began to read: Kandell, Damasio, Ramachandran, and the rest. Theology may have originally prompted my inquiry but my readings in neuroscience then doubled back to shape my theology. One notable example: My image of God is more like "mind" than like "brain." Both are verbs or adverbs—not at all nouns. In that same interview, Kandell mentioned that this would be the century of neuroscience, and he anticipated a variety of new fields, one of which would be neurotheology.

In the midst of this work, I received an e-mail from a Reform rabbi in Fort Worth, asking me if I would be available to discuss theology when he spent a sabbatical in New York during the summer. Ralph Mecklenburger probably anticipated that if I would be around, I would be more than eager to discuss theology, particularly when he mentioned that he had read my books and was very much interested in the work of theologian and Christian existentialist philosopher Paul Tillich, from which I drew some of my main ideas. So we met and continued to meet, summer after summer, during what he called his "study leaves" in Manhattan, and we talked theology.

In retrospect, it was probably inevitable that the interface between neuroscience and theology would eventually come up in our conversations. To my mind, the link was clear: Theology is a form of knowledge; the study of how we know *anything* is epistemology. Somehow, all of epistemology depends on the activity of the human brain. The forms of knowledge associated with religion/theology must then also depend on brain activity.

The systematized study of the activity of the human brain is the domain of neuroscience. It follows, then, that the central issues of philosophy are all implicitly epistemological, and all require an understanding of the workings of the human brain. In brief, it is inevitable that theology too would encounter neuroscience. This would indeed become then the (relatively) new science of neurotheology—on the crudest level, what goes on in my brain when I think of God.

I recall, somewhat painfully, posing that question to the scientist father of a student of mine. To my question, "How does a chemical reaction in my brain yield a concept of God?" His answer was, "Professor Gillman, that's all there is." I don't recall if I shared that exchange with Rabbi Mecklenburger at our next meeting, but I do recall hoping that his work would provide a more considered response to that question.

It has accomplished considerably more than that. It has, in fact, highlighted the neurological underpinnings of all forms of religious expression. Not simply neurotheology, it is an introduction to the discipline of neuroreligion. If, as it seems obvious, we carry our brains/minds (whatever the relationship) with us wherever we go, then we also carry them with us when we worship, when we perform religious rituals, when we seek community, when we behave morally, and when we contemplate what we mean when we use the term "God." There is no doubt that this entire range of issues will raise troubling questions about what we believe and how we live our religious traditions. The comment by my student's scientist father provided a powerful illustration why many religious people will do everything to avoid this entire field. But why do that? The issues will never go away.

Rabbi Mecklenburger's book will have accomplished its major purpose if it provokes us to become aware that there is a neurological dimension to all that we do. If this is to be the century of neuroscience, then we must jump in. And for those of us who need to do theology, then theology is where we jump in.

—Dr. Neil Gillman,
Simon H. Rifkind and Aaron Rabinowitz
Emeritus Professor of Jewish Philosophy,
The Jewish Theological Seminary of America

Introduction

A revolution in human self-understanding is under way. The dictionary I took to college in the 1960s defined "the humanities" as "the branches of learning concerned with human thought and relations, as distinguished from the sciences; especially, literature and philosophy, and, often, the fine arts, history, etc."[1] Today the fourth edition says the same thing.[2] But it should not. Cognitive studies reveal more every year about "human thought and relations," shedding light not only on medical issues but also on how and why we think and behave as we do. This in turn opens the way to deeper understanding of why we produce and enjoy the "literature and philosophy, and, often, the fine arts, history, etc." that mean so much to us. I took the required science courses in high school and college to get them out of the way and move on to the humanities in general as an undergraduate and to theological studies in particular as a graduate student. I aspired to know about beliefs, values, aesthetics, and what motivated individual and group behavior—what makes us tick—and what a modern person could believe about life, God, and our place in the universe. I would scarcely have imagined during my school days that I would one day produce a book intended to give science a voice in shaping religion and theology. But here it is. Cognitive science is dissolving the boundary between science and the humanities. All that we do and everything we care about are influenced by the way our brains operate. Religion, I have found, has much to learn from science.

Still, this is primarily a book about religion, not a neuroscience text. Although you should expect to find a significant dose of science in this book, it avoids most technical terminology and omits countless fascinating and important cognitive studies topics that did not strike me as of direct relevance to religion.

From the beginning of my rabbinic training, theology has been my passion. Not that I do not love the Bible, biblical interpretation, and other areas, but my rabbinic dissertation was in the realm of theology. Furthermore, within that discipline over the years I developed a special fondness for the field of interfaith dialogue, which I was privileged to teach for many years at Brite Divinity School. That, too, without my realizing it until recently, was preparing me for some of what follows in this book. I write, quite obviously, as a rabbi but have done my best to address a broad audience, considering Christianity, too, and occasionally other faiths.

As you will find at the beginning of chapter 1, a quarter of a century ago simple curiosity led me to read about consciousness and the brain in order to gain insight into our young son's attention deficit disorder (ADD).[3] As my fascination grew, I read more, and for the past dozen or so years a lot more, as the field of cognitive studies all but exploded with new research and insights, and as this book slowly took shape in my mind. My study long since ceased to have much to do with understanding our son and became an exploration of human nature in general.

We cannot understand our humanity without an appreciation of how our brains construct our reality and guide us through life, finding and assigning meaning in the process. I dare say every field we lump together as the humanities can be better understood with the help of cognitive studies or cognitive science, terms far broader than neuroscience, encompassing psychology, evolutionary biology, philosophy of science, and so on.

Many of the authors whose scientific insights excited me gave hints along the way that they realized the study of the brain had implications not only for science, clinical medicine, and even art, literature, education, investing, and so forth, but also for religion. Yet whether for fear of provoking controversy that might distract from their scientific work, or perhaps from a sense of academic inadequacy in the field, with rare exception they seemed hesitant to draw theological conclusions. As a serious reader of cognitive science with a deeper background in theology, I have attempted to draw such conclusions. Such a multidisciplinary approach is needed because, as a neuroscientist friend who read the manuscript said to me recently, "As 'war is too important to be left to the generals,' neuroscience is too important to be left to the neuroscientists." To the extent that I am successful here, my "value added" derives from elucidating what I think of as the intersections of cognitive science and theology.

We shall begin with a look at how consciousness is formed and why for the most part we must trust our own perceptions and mental processing. We know we make mistakes, yet we operate with faith in ourselves and faith in the consistency and predictability of our environment. This is not religious faith, I suggest, until we add ideas about the Divine, the topic to which we turn in chapter 2. Chapter 3 then asks what is happening in our brains when we have a religious experience and what spirituality, heretofore a very loosey-goosey term, actually is. Chapter 4 will suggest that in different eras Western religions have meant different things by "soul." Although the term is a useful and meaningful metaphor for aspects of consciousness, science undermines the classic idea that the soul is an add-on to the body without which we could not live or without which we could not appreciate ideas in general or intangibles such as beauty and love. That theological bombshell necessitates a digression into ideas of immortality. In chapters 5 and 6, we turn to free will and morality. Free will, I argue, is not an illusion but also not as absolute as many think. Even with free will—as the classic Christian idea of original sin and Rabbinic Jewish teachings on the evil inclination have long taught—we remain morally flawed creatures. Cognitive science can help us appreciate morality, as well, which is grounded in reality. Some moral notions actually appear to be hardwired in our brains. Rounding out the study, we turn in the final two chapters to what cognitive studies reveal about our hunger for community, which is one of the foundations of what has been called—sometimes derisively—organized religion, and to the great number of religions. While considering organized religion in chapter 7 we shall linger over ritual and prayer as well. In the final chapter we ponder the strange fact that there are not only many different religions but also multiple streams or denominations within the major ones. Modern understanding of the way our brains operate, turning experience into narrative and narrative into identity, provides an intriguing clue to what holds together these different families of religion. There is no one-size-fits-all religion, I suggest, so we should be respectful within and between families of religion, though to achieve full benefit we must be genuinely devoted to one.

Today, as always, there continues to be a vital role for religion, which grows out of the very structure of our brains and thinking and our individual and social needs. The brain and consciousness are themselves awe inspiring. So learning more about them no more undermines religion than learning about how symphonies and paintings are crafted takes away from our appreciation of music and art. Science alone does not provide the ultimate answers or firmly rooted values for which we yearn. But religion alone does not have all the answers, either. We are blessed, as moderns, with both. I should note that because cognitive science remains a young field with a great deal yet to be learned, major discoveries no doubt lie ahead, and some of the science I attempt to summarize and draw theological conclusions from may well need revision in years to come. Theology, too, will continue to evolve. But it would be cowardly on that basis—not to mention condemning religion to the status of a stale museum piece rather than a vibrant source of inspiration, guidance, community, and comfort—to refrain from drawing conclusions, some possibly controversial, based on the new understandings coming to light. My hope is that you will gain a greater understanding of our brains and consciousness and of the religions they shape.

ACKNOWLEDGMENTS

My wife, Ann, has spent hours researching sublets and making other arrangements as summers approached each year, facilitating the study leaves without which this book could not have been written. She has also never complained as I disappeared for hours at a time each weekend to write. As with so much else in my life, this book could not have been accomplished without her love and encouragement.

A dozen years ago the people of Beth-El Congregation in Fort Worth, Texas, generously offered me a couple of summer months for study every five years. After the second of those study leaves, I suggested that what I really needed was a month *each* summer. And it was so. Beth-El and Fort Worth have been a fine home for nearly three decades, and I am grateful for the friendship and encouragement of more people than I could list.

The first summer in New York I wrote in advance to Dr. Neil Gillman at the Jewish Theological Seminary. I had recently read his book, *The Death of Death: Resurrection and Immortality in Jewish Thought* (Jewish Lights). I was particularly intrigued by his use of Protestant theologian Paul Tillich's notion of "living in an age of broken myths" and wanted some further advice on readings. "You want to talk about my book?" he said with delight. "Of course!" Neil and I have become warm friends, meeting summer after summer and trading e-mails and phone calls in between. Furthermore, he read and critiqued each chapter of this book, questioning, challenging, and urging me forward. For me he has become the very embodiment of what a rabbi and teacher should be.

Particularly to avoid any technical misstatements about the brain, I sought out the help of my friend and congregant at Beth-El in Fort Worth, Dr. Howard Kelfer, a pediatric neurologist. He, too, read these chapters one by one as they were written, always getting back to me with remarkable alacrity and helpful suggestions. His enthusiasm for the project and his expertise are greatly appreciated.

Several others deserve acknowledgment. I wanted to be sure someone who knew a lot more than I about the Christian idea of original sin read my discussion of that classic doctrine. Dr. David Gouwens, professor of theology at Brite Divinity School in Fort Worth, graciously did so and provided important corrections and guidance. Once I had a first draft of the full book completed, two other readers made useful comments. Thus my thanks to Rabbi Elie Kaplan Spitz of Congregation B'nai Israel in Tustin, California, who has an interest in the brain and religion—and a neurologist spouse with whom to discuss it—and to Dr. Daniel S. Levine, a professor of psychology at the University of Texas at Arlington, whose writings include some work on neuroscience and religion.

My thanks go, too, to Emily Wichland, vice president of Editorial and Production at Jewish Lights Publishing/SkyLight Paths Publishing, who has been unfailingly pleasant and encouraging—and has an amazingly sharp editorial eye. Publisher Stuart M. Matlins also provided much appreciated counsel and encouragement throughout the process. The usual caveat goes with all these thanks and acknowledgments. This is a better book than it could otherwise have been because people helped, but responsibility for any flaws remains my own.

Finally, a couple of housekeeping details: biblical quotations, except as otherwise noted, are from the 1985 Jewish Publication Society translation of the *Tanakh* (Hebrew Bible). Among the materials at the back of the book are suggested readings and a selected bibliography of cognitive studies to assist readers who might like to dig further into the field.

1

Our Believing Brains
On Not Being Overwhelmed

Our brains construct our world in order to guide us safely through life, creating and appreciating meaning as we go. At the simplest level we must screen out a great deal of data or be overwhelmed, and even at more advanced and abstract levels there are always more options than we can evaluate. So we develop screens, "paradigms," and, based on what we are taught when growing up and our further experience and education, put our faith in them.

There is nothing either good or bad,
but thinking makes it so....
William Shakespeare, *Hamlet*, act 2, scene 2

I started reading consciousness studies not long after my son was diagnosed with attention deficit disorder. Alan had always seemed normal and likable to me, but I had to admit

1

he was not doing well in school. The psychologist who worked him up both frightened and reassured us. Tests designed to measure the cognitive functions of our little kindergartner were all over the map. In some areas he showed extraordinary capabilities, and in others he was way behind what was expected of children his age. With such a wide variety of capabilities, from way behind to way ahead of the norm, the doctor told us, "If this were a physical rather than a mental disability, he would be in a wheelchair." Wheelchair? My cute little boy with the Paul Newman eyes that the ladies in the synagogue were agog over? That was the scary part. The reassuring part, however, was that he had wonderful verbal skills, which would be particularly important in the adult social and work world. The challenge was going to be getting him through school, with its wide variety of challenges. As an adult he should be fine.

We now know the wheelchair image was hyperbole. Alan, it turned out, needed a special school that could adapt to his strengths and weaknesses as a student and teach him to adapt to them. Hill School insisted parents think of "learning differences, not disabilities." By high school Alan was ready to be "mainstreamed." He graduated high school with decent grades and fabulous SAT scores, including a perfect 800 in mathematics. Five years later he graduated business school at the University of Texas and got a job in computer consulting, and recently he married. He is fine.

But my wife and I could not know that future when the psychologist presented the case to us as that of a mental paraplegic. I wanted to understand not merely the significance of the fact that Alan was easily distracted, internally as well as externally, but also what that really meant. How was his brain any different from mine? So, first with Daniel Dennett's *Consciousness Explained* and later with lots more books, I began reading about consciousness, brain function, and evolutionary biology. I learned early on that one of the chief things our brains do is take in huge quantities of data from all our senses and then, in effect, filter out what we do not need at the moment to focus on what may be useful. If you are sitting in a chair reading this paragraph, for instance, prior to my mentioning it you were probably unaware of the feeling of the chair on your bottom and back or any number of sounds and sights around you (as I wrote that, I noticed the air-conditioning system humming in the background). Those perceptions were available to you if you needed them, but you were concentrating on your reading. "Attention deficit," I came to think, was actually a misnomer. My son's problem was an attention *surplus*. He was paying attention to too many things when he needed to concentrate on one. And especially at age five or six he had no control over that. Do you or I have control over that? Not total control, but some. If you are easily distracted by movements and noises, you should study in a quiet, relatively isolated place, and you certainly should not play music while reading. But if, like me, you become so engrossed in reading or writing that you scarcely hear music or notice people walking past in the library or your office, that will not be necessary. As they said at Hill School, this is not a matter of the right or wrong way to work, but simply of differences in the way each of our brains is wired.

What I labeled "attention surplus" led, of course, to a deficit of attention to the specific lesson my young son needed to concentrate on. The nomenclature is less the point than understanding how the brain works. With a chemical assist from the drug Ritalin, trained teachers, and his own native intelligence, Alan learned how to manage with his deficits, just as each of us—because no one is perfect—learns to adjust to our deficits. I, for instance, long ago realized that I never heard a name I couldn't forget! Although names are still not my strong point, I have learned to pay better attention, repeat them, associate them, and thus remember more of them. Alan, I later learned, also has strengths to maximize that, for general success in the world, help make up for his weaknesses. He seems to have an intuitive feel, which I totally lack, for the workings of machines, including computers (my father was an engineer; it just skipped a generation). Each of us needs to minimize our weaknesses and maximize our strengths.

So I started reading about the brain and the mind in response to a specific concern. But soon I was hooked, and for over twenty years, as the fields of consciousness studies and cognitive neuroscience burgeoned, I have kept reading. Our minds can only do what our brains are wired for. This is a revolutionary new area of science in which more is discovered every year—which also means yesterday's understanding is constantly being challenged and refined. To a remarkable extent we are our minds; we know nothing but what makes its way into our minds. So the way our minds, and the brains that underlie them, operate has immense importance for every field of study.

Including religion. The more I learn about the brain, the more I realize that the way it works has tremendous relevance to the questions that have most interested and sometimes troubled me as a rabbi. Our beliefs, our spirituality, our sense of community, our relationship to people and God are no less dependent upon our brains than are reading, laughter, sports, problem solving, love ... and everything else we do.

I am a rabbi, not a neuroscientist. So I shall avoid technical medical jargon. What I will not avoid, whether they seem to confirm traditional beliefs or to throw them into doubt, are the theological implications I have found in pondering this new scientific frontier.

WE CONSTRUCT OUR REALITY

Our brains do a lot more, of course, than screen what would otherwise be overwhelming quantities of data coming in from our senses. First of all, they process the data and focus our attention on what seems most compelling. The processing is remarkable. When you see a pillow, for instance, your brain notes its shape, its color, its texture, and more, comparing it to similar things you have seen sitting on couches or beds in the past so that you will know what to call it and what to do with it. When you hear a trumpet, you perceive not only the pitch—the musical note—but also the timbre (a C on the trumpet sounds different from a C on a violin or piano), and then you further discern the tune and its rhythm. These different aspects of our perception are processed in different sites in the brain and then put together so that we experience them as a whole—the plush red couch pillow or the Purcell Trumpet Voluntary. With each of our senses, similarly, we receive data and reconstruct it into perceptions that we can use. It is not an exaggeration to say that our brains construct our reality constantly. Before your brain decides to ignore the siren going by outside and focus on the pillow, or vice versa, it is taking discrete aspects of such stimuli and putting them back together for your recognition (that squarish, fluffy red thing is not a giant red marshmallow; it is a pillow—though if you saw it floating in a huge cup of hot chocolate you might reach the opposite conclusion, so your memories are involved, not your perceptions alone).

The brain does this so quickly that it seems instantaneous. Some processing shortcuts help. We need not pay careful attention to the details of each tree we come across. We are used to seeing trees, and the brain assumes a typical oak or pine (without, of course, needing to know the name), unless for some reason we decide to pay added attention to the pattern of limbs, leaves, or needles. When we see a checkerboard pattern tile floor or wallpaper design, we assume consistency for the whole floor or wall, which on closer examination is usually, but not always, correct. Again, the point is that the brain processes the data and constructs our experience. When two things occur in rapid succession—you drop a book and hear a thud—you are apt to assume the former caused the latter. This is so often correct that our brains save effort by assuming it. Occasionally, though, we become aware that we were mistaken (the magician puts his hand to someone's ear and a penny suddenly appears, then covers the coin with a cloth, waves his magic wand, and it disappears; we perceive that the hand pulled the penny from the ear, and the wand made it disappear, but actually it was all sleight of hand and habit of mind).

While all this and more is going on related to our perception of external reality, simultaneously our brains are tracking and helping control internal physiological systems. In my car, a gauge monitors the gas tank and shows me when fuel is running low. In my body, the brain monitors my digestive system, and various chemical levels here and there, and subtly makes me feel hungry or thirsty. There are also temperature, respiration, and other processes to watch and adjust. Muscles must be activated and coordinated if I am to walk, run, or swing a baseball bat—and to hit the ball or duck a beanball I'd better be able to coordinate vision and muscle action. The brain multitasks all the time. It also constructs internal maps of rooms, buildings, neighborhoods, and whole areas that I have navigated before and may wish to get around in again. Further, there are people to deal with. Do their faces look friendly? Have I seen these individuals before? How shall I make a good impression on them, scare them off, or do whatever it is that I wish to do?

We store great quantities of information and constantly need to make decisions. Little of this is at the level of abstraction we generally think of as higher thought—reading a book, writing a letter, solving a math problem, or plotting strategy for a game, a business deal, or a seduction. We learn to do that, too. Very importantly, an emotional system parallels—and impacts—the basic perceptions and the more abstractly rational system. We pay a lot better attention and remember more accurately when something is alluring or scary, exciting or depressing.

Specific areas of our brains are involved with each of these tasks and many more. The wonder is that it all comes together. We feel as if we are experiencing reality whole, not fragmented. Our brains are constructing the fragments into the whole. There is not even some one site in the brain where it all comes together. As the different brain regions operate and communicate with one another (relatively) simultaneously, we have the subjective experience of being selves in a world.

The current moment usually makes sense to us because our brains have experienced trees, chairs, living rooms, forests, and highways, not to mention recurring situations—going to the store, embracing a loved one—before. Furthermore, we do not feel as if we are living a bunch of discrete moments but experiencing a flow of seemingly related events. In a word, an area of our brains is charged with making a story, an ongoing narration, of our lives not only year to year, but even moment to moment.

We could go on and on with all this. If it interests you—and I find it fascinating, for it is nothing less than the description of how I operate, the foundation for who I am—I suggest you read one of the many books that have come out in recent years on how it all happens (see suggested readings for chapter 1). My purpose here, as background for a consideration of intersections between neuroscience and religion, is to stress that we are not simply perceiving exactly what is out there in the world. We receive data and put it together, influenced by our previous experience, and often adding affect as we reconstruct it all into our experience of living. For the most part, we trust our experience. But we know we do not get things 100 percent right all the time in the first place, much less remember things perfectly later.

Although much remains to be discovered about brain functioning, an amazing amount has been learned about the way our brains—each made up of a hundred billion or so neurons and another trillion glial cells, which "do the housekeeping" and provide the proper environment for the neurons to do their work of electrical and chemical signaling[1]—perceive the world and guide us through it, usually safely. At first there may seem little directly to do with religion here, at least beyond a sense of awe as we realize how intricate we are. But of course all of this has tremendous relevance for understanding everything we do, because we *are* in large measure our brains (also the rest of our bodies, of course, but we only know our bodies via our brains). So it is time to build on the observation with which I began. The amount of data coming into our brains is so great that we are incapable of paying attention to all of it at the same time. Our brains screen and sort, focus and ignore, *and try to make sense of* where we are, what is happening to us, why it is happening, and what if anything we ought to do about it. With the effort to make sense of our experience, we move into the realm of meaning, a realm of deep concern to religion.

WHY WE MUST BELIEVE

You have no trouble walking through an odd-shaped room in your home, stumbling neither over the furniture that is always there nor the toys that the children have left strewn about. Your mind has mapped the space, and your vision guides you to step safely over or around the obstacles. If you were blind, you could do almost as well. The toys would be a problem, though if you knew they were often left out you could feel ahead carefully with your foot or with a cane. We trust our senses. Occasionally they mislead us, but not very often. We trust our memories, similarly imperfect but plenty good enough for most daily tasks.

When you turn on the tap to brush your teeth, you do not check if the water is safe, nor when you cross that narrow bridge from the parking structure to the office building do you wonder if it will hold your weight. Not that your perceptions have been switched off. If the water ran brown, you would notice. If there were a jagged crack in the concrete, you would probably not put your weight on the bridge. Lacking such signs, it is possible in theory that the water quality has been compromised overnight or that the bridge was poorly engineered and, though it looks fine, is about to collapse. But the probability is so slight that you do not worry about it or even think about it.

We could not function—we would be afraid to get out of bed in the morning!—without trusting our senses, our memories, and the consistency we find in our environment (the water system, the bridge). This despite the fact that if we stop to think about it, we know that occasionally we misperceive, remember incorrectly, or experience some unforeseeable inconsistency in the environment. This can even be catastrophic, as when a semi suddenly crosses the median on the freeway or an earthquake suddenly renders terra firma not so firm. But our brains, with the input of our senses and the benefit of our experience, get us through day by day remarkably well most of the time. We have to believe our senses. We must trust our memories. We count on the consistency and predictability of our environment. In a word, we live by faith.

I am not suggesting that we must believe in God, merely that we must act, virtually all the time, as if the world we perceive and construct is benign. It is possible to have the water tested, and the bridge inspected, and it is wise to pick up the toys. But when we realize everything that could go wrong not merely day to day, but also minute to minute, we have no choice but to believe—beyond the evidence—that perceptions, memories, and environment are reliable.

I should add here that though theologians can make worthwhile distinctions between belief and faith, I am here using them in the everyday sense as synonyms for that which we accept without requiring proof. We would be overwhelmed not only if we had to pay attention to all the data coming at us constantly, but also if we had to worry about the accuracy of what we are getting (externally) and considering (internally). When we notice something unexpected, we can quickly alter our actions (don't drink that brown water; step back from the cracked bridge; swerve away from the oncoming semi). But for the most part we must believe that our reality is all right, or we would be overwhelmed with worries and the attempt to foresee unlikely contingencies.

We have moved from the brain correcting for overwhelming data to the brain trusting its own construction of the world and the prior experience that helps us evaluate it. Dr. V. S. Ramachandran, a neurologist at the University of California at San Diego, presents several incidents of Capgras syndrome. In his book *A Brief Tour of Human Consciousness*, a man in a terrible car accident wakes up from a coma and seems quite normal, with the bizarre exception that when he sees his mother he insists she is an impostor, and no one can convince him otherwise. She looks like his mother and talks like his mother, everyone reassures him she is his mother, and on the phone he acknowledges her as his mother. But whenever he sees her he insists she is an impostor.[2] Another patient with this fortunately rare condition, when pressed for an explanation of why the person, in this case his father, would pretend, responds, "That is what is so surprising, doctor. Why should anyone want to pretend to be my father? ... Maybe my real father employed him to take care of me, paid him some money so that he would pay my bills."[3] Even when faced with the witness of others and compelling arguments that the person is not an impostor, the Capgras patient will not relent. The actual explanation, Dr. Ramachandran says, involves the way our brains work:

> To understand this disorder, you have to first realize that vision is not a simple process. When you open your eyes in the morning, it's all out there in front of you, and so it's easy to assume that vision is effortless

and instantaneous. But in fact within each eyeball, all you have is a tiny, distorted upside-down image of the world. This excites the photoreceptors in the retina and the messages then go through the optic nerve to the back of your brain, where they are analyzed in thirty different visual areas. Only after that do you begin to finally identify what you're looking at.... Finally, once the image is recognized, the message is relayed to the amygdala, sometimes called the gateway to the limbic system, the emotional core of your brain, which allows you to gauge the emotional significance of what you are looking at.[4]

Capgras patients, like the rest of us, see someone who looks like a loved one. But the link between the analysis part of the brain and the emotional part of the brain has been damaged in the Capgras patients. They do not experience the familiar feeling, the emotion, they got before when seeing their mother or father (also measurable as galvanic skin response—a bit of sweat). Because they don't have that feeling, they "know" the person is an impostor.

But why can the Capgras patient not simply recognize that all these people who seem to mean so well are telling him the truth? If this were simply a logical puzzle, we might answer, "Because if I do not trust this perception, why would I trust any other?" But that logical step is not there. We trust our construction of the world because it *is* our construction of the world. We evolved that way. We are hardwired that way. We can learn to do something counterintuitive, say steering into the skid as we lose control of the car on an icy road, but it will still feel like the wrong thing to do. Add the emotional component of not recognizing a parent because she just doesn't feel authentic, and we can learn to speak to her in a kindly way and call her "Mom," but we will not really believe it is her. Dr. Ramachandran and others have described lots of these cases, not only Capgras syndrome, in which damage to a part of the brain leads to seemingly crazy behavior. If one of the brain areas involved in face recognition is damaged by accident or disease, for instance, you may get phenomena like Dr. Oliver Sacks's memorable book title *The Man Who Mistook His Wife for a Hat*.[5] What seems unnatural to the observer does not to the individual, who is, after all, simply believing his own brain's construction of the world, as each of us must.

That most of us do not exhibit such bizarre behaviors testifies to the fact that for normal, everyday purposes, our perceptions are good enough and our mental processing of them, though incredibly intricate, is remarkably efficient. Our brains construct a world that we can, and indeed must, believe in. We do not believe this because we are always right; we are not. Yet we are usually right, and (as discussed before—the water in the tap, the bridge) we would be immobilized if we did not trust our gut feelings.

I do not want to get to God until the next chapter, but it is worth pausing again to say we live by faith in the nontheological sense of believing what goes beyond the evidence but does not contradict the evidence. Our minds, at this still rather basic level—before culture and religion, just navigating through the world—*need* to act as if there were basis for faith in ourselves and faith in our environment. We know we make mistakes interpreting our sense perceptions. And we know the world does not always behave the way we expect it to behave. Yet we not only act as if we are confident; we genuinely feel confident. Without that faith we would be overwhelmed by uncertainty.

OUR PATTERNS OF BEHAVIOR, OUR PATTERNS OF MEANING

After waking in the morning, usually I wash and dress, go to the gym to exercise, shower and dress again, then have breakfast and go to my office. This is a morning ritual, a habit. Parts of my routine are healthy, but the order is neither good nor bad, just comfortable for me. Doing this as habit frees me from deciding what to do. My work is so varied that it cannot be as routinized. But when nothing unusual interrupts, once arriving at my office I check for phone messages, then e-mail, and later look for other mail. At midday I go out for lunch and late in the afternoon I go home for dinner. There are lots of things to be decided along the way—which messages to answer right away, which to ignore entirely or deal with later; when to write a sermon, make a hospital call, put together next month's bulletin, drop everything and handle an emergency, and so on. That some of this is routine helps me cope with the rest. Although not everyone is equally a creature of habit, we each have our routines. This is another mental strategy to avoid being overwhelmed by virtually infinite possibilities.

The pattern of weeks and weekends, months and years, holidays and vacations similarly creates a certain rhythm to my life. If I moved to a different society or adopted a different religion, some of this would change, but the fact and utility of routine would probably not change much. Furthermore, culturally based value judgments, such as the importance of work or the holiness of Sabbaths and holidays—or football games—strengthen our routines. We begin to invest emotionally in them and find them not only useful but also meaningful and satisfying.

Even before such cultural patterns call forth emotional investment, interpersonal relationships do. As infants we cried, and someone tended to our needs, and probably cooed and cuddled with us, too. We developed a sense of mutuality and caring, of love, with parents and then family, and eventually with friends. We were warned what not to do and what we should feel obligated to do, and rewarded or punished. All the while we were learning what did and did not work for our well-being, and eventually for the well-being of larger groups of which we were part, what could lead to our physical or emotional pleasure and satisfaction, and what, to the contrary, would cause pain, frustration, and suffering. Long before we could vocalize abstract notions, in other words, we learned patterns of behavior that we would be taught to call good and bad. We learned which could get us in trouble, or physically hurt, or emotionally shamed, or rewarded with something we liked—a smile, a hug, a piece of candy, cheers. Somewhere along the line we no doubt learned that sometimes we got away with more than we deserved or suffered unjustly.

We shall return in a later chapter to the question of whether morality is mostly learned or hardwired. For the moment, the point is that we deal with the infinite range of ways to live and attitudes to have by developing—routinizing, if you will—thought patterns. A day is a day like any other until we learn about cycles of weeks, months, seasons, years, and so on. A place is a place until we learn one or another place is special—home, homeland, or perhaps office—and begin regarding other places vis-à-vis the special places. We are taught rules for other activities of our mind: grammar and phonics, mathematics, geography, and on and on. Individually, and as the species *Homo sapiens*, our minds crave order and meaning. We try to discern pattern and significance or, failing that, to create them. We can readily see how cultural customs can and do change—the daily routines, the cuisines, the clothing, or holidays—but most of us do not like familiar patterns to change too much or too often, for these, too, help our finite minds cope with infinite possibilities.

Our minds are exquisitely tuned to find and create patterns and variations. A key part of the architecture of our minds, moreover, is the ability to connect things and events over time. As mentioned earlier, we do not experience the world or our lives anew moment by moment—a series of freeze-frames—but sense, rather, a flow of events shaped in large measure by cause and effect. What is more, often we come to see our personal life stories as part of larger histories—our family's tale, national history, religious saga.

People love stories. Why? Because, as mentioned, we turn our own experience into narration. Our brains are structured to make connections. We want to sense how one experience leads to another, what connects our lives into coherent wholes. I started here, then that happened, he did this and I did that, and it led to something else. Stories with a beginning, a middle, and an end are especially neat and comfortable. It is not always so clear what the pattern is in our lives, or in some smaller aspect of experience, so—sometimes consciously and sometimes not—we search for that pattern. There is satisfaction in figuring out mysteries, recognizing characters' motivations, whether in real life or in fiction. To best deal with life, taking advantage of opportunities and avoiding dangers, we naturally want to understand what is going on about us and why.

Most of us, most of the time, manage reasonably well. We experience the world and what is going on, and a set of brain circuits comes up with a reasonable explanation. Dr. Michael Gazzaniga of the University of California, Santa Barbara, speaks of fitting our experience into preexisting schema. We see something going on, and the brain, based on previous experience and attitudes, explains it. According to Dr. Gazzaniga, "Our brain is adapted for extreme efficiency; for that reason it distorts incoming information to fit in with our current beliefs about the world."[6] When we speak of areas of the brain doing discrete tasks—mapping a room, seeing a shape or texture, monitoring a chemical level—we mean literally specialized locations in the brain. To treat certain extreme epilepsy symptoms, it is sometimes necessary to sever the neural pathways that connect the left side of the brain to the right. Dr. Gazzaniga and his colleagues have experimented with such patients in an attempt to learn about the different functions performed by each side. He writes:

> In the study that most vividly demonstrated the left brain's ability and need to interpret information, I presented a picture of a house in the winter blanketed with snow to the right side of the subject's brain, and a picture of a chicken claw to the left side of his brain. Remember that neither side of the brain knew what the other one saw. Then I asked the patient to choose one object with each hand that most closely matched the scene he saw. The patient chose a rooster with his right hand, which is largely controlled by the *left* side of the brain, to match the picture of the chicken claw his left brain saw. He chose a snow shovel with his left hand, which is largely controlled by the *right* side of the brain. As the patient did this, he was confronted with the fact that each hand had chosen a different object. Therefore his left brain (which processes language and deals with constructing verbal information, but never saw the picture of the snowy house) offered an explanation: he must have chosen the shovel because it could be used to clean out the chicken coop.[7]

In another experiment, the word "walk" was presented only to the right side of a patient's brain, leading him to get up and walk. He was asked why he was walking, and his left side's interpretation was, "I wanted to go get a Coke."[8] The implication, of course, is that whatever happens to us we unconsciously make sense of—most of the time correctly, but sometimes wrongly. Then, as we saw with the Capgras patients, we absolutely believe our own brain's conclusion. After all, imagine what a pickle you would be in—and some of these patients with malfunctioning brains *are* in—if you could not believe yourself, not believe your own brain.

But the significance goes beyond that. The man who chose the rooster and the shovel and then unconsciously came up with an explanation was creating for himself, as well as explaining to the researcher, the meaning of his action. In normally functioning brains we probably get our explanations right far more often than wrong, but the point is we are assigning the meaning. *We not only construct our world, but we also assign its meaning!* (And, not to put too fine a point on this, it is worth saying that once the shovel is for manure or the walk is to get a Coke, it is not quite fair to say those are "wrong" explanations. The subjects' minds could have fabricated different explanations, but once these were stated, they genuinely became, for these individuals, the meaning. To get us through life, our brains have to trust themselves.)

CULTURAL PARADIGMS HELP US COPE

So much data! So many ways to construe it! So many action choices! So many ways to explain those choices and fit ourselves into the constantly evolving picture! Not to be overwhelmed—that is our mantra in this chapter—we focus, interpret, narrate, assign meaning, and still find ourselves facing unfamiliar situations and driven by urges of which we have no innate understanding, desiring food, sex, power, security, love, and more. Life still threatens to overwhelm us. Fortunately, however, many of the situations that are new to the individual are familiar to the community. You, for instance, may not have had a baby before, but many have been born in your family, and far more in your community. You do not have to figure it all out on your own—what to expect physically, what to buy to be prepared, how to interpret this (here's that word again) overwhelming new set of possibilities. Family and friends step in. Religious rituals (depending on your tradition, namings, circumcisions, baptisms, and so forth) and secular rituals (name books, baby showers) and predetermined meanings are offered. Unlike the nearly absolute need to accept our own sense data and mental construction of reality, we have some choice about whether or not to accept what the community tells us in these situations. In fact, though, the possibilities for what to do and how to regard the situation—the meaning—are so varied that we generally welcome the culturally proffered help. Whether from family, religion, regional or national folkways, or elsewhere, paradigms exist for fitting what is new experience to us into tried-and-true categories. Life is full of new challenges, but we do not often have to face them totally on our own, to reinvent the wheel. This applies most obviously to major life turning points such as birth, puberty, marriage, and grief. But it applies as well to an endless array of situations, attitudes, and customs regarding child rearing, education, courtship, work and careers, and so on—even entertainment, as we become loyal to certain sports, teams, or styles of music. This is not to suggest that our religion, nation, or any other cultural grouping has absolute power to program our responses. Still, specifically because there are so many ways to understand life and the world, so many options to choose from, if left to our own devices we could be confused much of the time. So giving our loyalty to one or another of them—most often the one parents urged upon us—is another very helpful coping mechanism for our brains. This is the level at which big questions intrude upon our awareness: Why are we here? How, as individuals, shall we live? How shall we organize society? What is life's meaning?

Most of us would consider being open-minded a virtue. You are, say, a Republican or Democrat, but you want to know what members of the other party think, and not just to plot how to overcome them, but because they, too, sometimes have good ideas. Socialists may appreciate the power of market capitalism to foster rapid economic growth, or capitalists see the utility of collectivist solutions for a society's defense or health needs. The avowed atheist may find an uplifting idea in the prophecy of Amos or Isaiah, and the Jew may appreciate the sayings of Lao Tzu or the marvelous beauty of a Christian cathedral. Even for those anxious to keep open minds, though, the sheer volume of competing traditions and cultural paradigms about everything is so vast that we need to screen some out to focus on others. I am far more apt to read a new book on Judaism than on Shinto and certain to ignore the steady stream of publications from what I generally regard as cults (though I cannot possibly know very much about most of them). If someone I respect, or a book review digest, pitches a book to me that I would otherwise ignore, open-mindedness may lead me to take a look at it; however, it is literally impossible to be so open-minded as not to use our developed loyalties to screen out a great deal, indeed most, of what is available.

To put all of this differently, with first parental, and then community, input, we develop loyalties to these large paradigms that help us interpret life. Whether through childhood indoctrination and conditioning, mature decision making, or probably some combination of factors, we all do this. We must. We must believe in something (or several somethings) not to drown in all the possibilities. If we rebel against what parents and teachers provide us, we will find something else. Remember: we construct our world and we assign its meaning, and the potential actions and interpretations are dizzyingly manifold. Not choosing is not an option. The brain craves patterns and connections, creates meaning. That is who we are, what our brains, at all levels including this highest level, do. The issue is not *whether* we will select, privileging some ways of interpreting over others. The issue is only *what* ways of thinking we shall give our allegiance to.

RELIGIONS: SCREENING MEANINGS IN AND OUT

Many of the largest of these paradigms that we use to screen and organize our experience are religions, which usually refer to God or gods, or at least some realm of the holy, to help make sense of and posit a meaning for our journey through the vast and potentially overwhelming world that we need to navigate. But there are plenty of lesser paradigms of considerable significance. Trying to explain this to a physician in my synagogue, for example, I suggested that as an MD he generally excludes witch-doctoring and folk medicines of many sorts from his personal reading and continuing education. He can see that there might be some value in folk remedies, and a placebo effect if nothing more in certain magical medical efforts, but even if there might be some value in these approaches, he has learned to place his trust in more scientific medicine. He has to have a basis for screening in what he most trusts and screening other things out, for there simply is not enough time to know about everything. The possibilities are seemingly infinite, and the time for dealing with them all too finite. Likewise, as a Jew, he cannot know everything even about his own tradition, much less devote his time to every other religious system he hears about.

This is not an argument for closed-mindedness. Occasionally, most of us appreciate having our horizons broadened, and on much rarer occasions we may even give up a paradigm because a new one appears more compelling. But especially for the biggest issues—Why are we here? How shall we behave? How do we deal with feelings of guilt and with the awareness and perhaps fear of death?—there are so many conflicting answers to the ultimate questions that, once again, we would be overwhelmed if we did not find the answer or set of answers that seems to work for us. Religion, the largest of the many frames or screens, is ubiquitous in human cultures, I suggest, because just as we would be overwhelmed at lower levels of our consciousness without an ability to screen, so too would we be overwhelmed without a set of overarching ideas, a structure of meaning in which to situate our life's journey. We crave the answers that religions offer. The values that they inculcate give us guidance and direction. This is first a matter of need and only then becomes a question of truth. Almost as much as Capgras syndrome sufferers must believe their own brains, committed believers (whether in a religion or some other structure of meaning such as a state or philosophy) are loath to abandon their faith, screen, and paradigm.

At each level of thought, then, from the most basic perception, to constructing our reality, to finding our way through the world and placing our lives in a context of meaning, we screen, focus, and find or create mental shortcuts. Without these mental strategies we would be confused and overwhelmed. We will believe in something. But will that something, in any given case, be worthy of our loyalty? Useful is not the same as true. The issue for theology, then, becomes: because we inherit as much as create religions, do they refer to something—or Someone—real, something "out there" or just a mental structure "in here"? What are our brains reacting to when we speak of God and religious experience?

2

Taking God Personally

Although no one fully understands God, awe in the face of our vast and impressive world and universe is a universal human feeling. Cognitive studies explain that our brains are designed to act as if everything we encounter involves us. Might that awesome reality out there be addressing us individually? There are major problems with personal-God theologies, especially explaining "why the good suffer." Personal-God theism persists anyway because of the structure of our brains: we naturally think everything is, personally, "about me." Order in the universe is undoubtedly real. Whether that Order personally relates to each of us is a question of faith.

The heavens declare the glory of God,
the sky proclaims His handiwork.

Psalm 19:2

A man said to the universe:
"Sir, I exist!"
"However," replied the universe,
"The fact has not created in me
A sense of obligation."

Stephen Crane

RELIGIOUS AWE AND THEOLOGICAL PROBLEMS

Get away from the ubiquitous artificial light of our modern cities, and on a clear night the star-spangled sky will take your breath away. These were the heavens that awed our ancient ancestors. Equally marvelous, though, are sights that only we moderns know, the intricate structure of creatures, crystals, and molecules, made visible only through microscopes. The beauties of nature at our human scale of perception, whether the single perfect rose or forest-clad mountain ranges, are nothing to scoff at, either. We humans, too, create things that fill us with wonder, from masterpieces of art and literature to the penetrating insights of philosophers and scholars. Deeds, certainly the bravery of heroes and martyrs, sometimes the performances of actors and athletes, and not uncommonly the devotion and determination of ordinary folks to persevere and overcome obstacles, even to sacrifice their lives for some greater good, can be awe inspiring as well. Ancient poets rhapsodized over the grandeurs of nature. The sensitive reader of the Hebrew Bible, whether taking literally the image of God as the universe's Creator or not, cannot help but resonate to the poetry of the book of Job as God speaks from the whirlwind declaring creation too wondrous for human understanding. The psalmist, too, feels humbled by nature, yet manages to affirm human greatness:

When I behold Your heavens, the work of Your fingers,
the moon and stars that You set in place,
what is man that You have been mindful of him,
mortal man that You have taken note of him,
that You have made him little less than divine,
and adorned him with glory and majesty?

(PSALM 8:4–6)

Modern religious thinkers such as Abraham Joshua Heschel try to tease out differences among experiences of the sublime, the wondrous, and the radically amazing and speak of our sense of awe and of the ineffable.[1] Rudolf Otto, a generation before Heschel, similarly spoke of different aspects of "the numinous."[2] Thus from the psalmists and prophets of old to religious thinkers of our own time, our sense of wonder in the face of that which is beyond us and not of our making touches us at the deepest levels of consciousness. This is a key part of our religious sensibility.

So when modern critics try to reduce religion to our fear of death and concomitant need for a Divine Protector,[3] or to Freud's mass neurosis or Marx's opiate of the people, I wonder how these people could be missing the *positive* impetus to religion? Are they blind to the Milky Way, deaf to birdsongs and symphonies, insensitive to love? Of course not. They are simply unwilling to apply the word "God" to this experience. That Power beyond us that we perceive in so many situations has been invoked to comfort and inspire, but also, all too often, to keep people terrified and in line and to justify intolerance and bloodshed. Moreover, while the gods or God may in the past have been needed to explain (or really cover for our lack of explanation for) how the world, and we ourselves, came to be and operate—the so-called God of the gaps—we understand a great deal more today and are confident of our ability to learn still more. In that sense, God may seem like an unnecessary or even harmful hypothesis.

Worse, basic theism—the understanding of God as Father/
King/Creator/Judge, like a Superperson except without a body,
and thus perfect and immortal, the "Lord of all who reigns
supreme," all-powerful, all-knowing, all-good—has always had
an Achilles' heel, the problem of evil. Already in the Hebrew
Bible the author of Job offers a theodicy, a defense of divine
justice. The prose prologue and epilogue suggest that when
we suffer undeservedly, God is testing us and will reward us
later. The main poem, more deeply, declares our merely human
understanding so inferior to God's that we simply cannot
understand divine ways. There have been many more of these
efforts to explain how a God who is like a person, and omniscient, omnipotent, and omnibenevolent, would allow evil. The
cleverest is the idea that reward and punishment, thus justice,
which often seems lacking in this world, will be dispensed
in another realm after we die, an explanation conveniently
unverifiable this side of the grave.

All those positive aspects of our experience that evoke our
awe, wonder, and gratitude—how fortunate we are to inhabit
such a world and universe!—remain real enough. But modernity
has deeply challenged classic theism as an explanation for them.
It is not that theodicy or science is new or that earlier generations
did not come up with explanations that are at least arguable.
One can neither prove nor disprove theism—or atheism, either.
Philosopher of science Thomas Kuhn's understanding of how
scientific revolutions work strikes me as relevant to religion
as well. When evidence arises throwing into question a basic
scientific understanding, at first explanations are sought, and
minor theoretical adjustments made. Even scientists are loath to
simply throw out an approach that works in so many instances,
particularly when they have no obvious substitute. But eventually all the tinkering leads to a general sense of dissatisfaction,
and a new theoretical framework is sought that can still explain
what the old one worked for but can also account for the old

approach's failings. Kuhn calls this a new paradigm.[4] In just this way, I suggest, the problems of classic theism have been seen and struggled with and the theory (or really theories, theologies) tinkered with endlessly. Ultimately, new paradigms are offered, not because classic theism is wholly untenable, but because it just no longer compels assent.

MULTIPLE METAPHORS FOR ONE GOD

Fortunately, Judaism, Christianity, and Islam, and Eastern religions, too, have long offered alternative ways to think about God. It would take us too far afield to go into depth about the numerous personal-God or what I shall call philosophical-God theologies. Granting, though, that it is merely a heuristic simplification, we can say that there is a set of theologies that conceive of God as being personlike and another set that conceive of God as a force or philosophical principle. The Hebrew Bible may appear to preach naive theism, taught the twelfth-century Aristotelian Maimonides, for instance, but that is simplistic surface-level meaning for the masses. Sophisticates see through all the anthropomorphism and storytelling to understand scripture philosophically. God is pure thought ever thinking itself and spinning (emanating) a world in the process. The Platonist Philo had approached scripture similarly eleven centuries or so before. Baruch Spinoza, in seventeenth-century Holland, spoke of God as the mathematical perfection of the universe, thus insisting that God and God's world could be understood in purely rational terms. The mystics of high Kabbalah, albeit with a radically different style, presented God as a collection of interacting spiritual forces. Some modern thinkers, such as Christianity's process theologians and Judaism's Mordecai Kaplan, speak of God as a process, not a being—in Kaplan's words, a "process working for good." For both rationalists and mystics the point is to make sense of experience, including, of course, what I have been speaking of as religious awe, and to continue to find relevance in Torah, even as they find personal-God theology difficult to maintain. Throwing all this into one paragraph necessitates omitting more than I am including—no pantheism, no existentialism, and so forth. I simply want to underline the distinction between these two styles of thinking about God. We are all reacting in awe to existence. Some are saying God *created* the order that awes us, and some are saying God *is*, in some sense, that Order.

I hold no brief for any of these particular philosophical-God ideas. Indeed, as a child of a scientific culture, I am going to suggest another. I mention them to demonstrate that more abstract ideas of God have been part of Jewish—and other—traditions for a couple of thousand years. Critics of such thinking complain that Maimonides and others forsook the God of Abraham, Isaac, and Jacob for the God of the philosophers. Similarly, critics of early Reform Judaism scoff at the 1885 Pittsburgh Platform's lauding of "the God idea" rather than the personal God. Traditionalists insist they are preserving the faith. But which version of the faith, which has never been monolithic? Preaching that which violates our era's notion of common sense is not a promising way of passing our religions on to the next generation, especially when the better schools teach our children science and critical thinking. That is why Maimonides was trying to guide perplexed Aristotelian-trained students to see that the best thinking of his era was compatible with biblical and Rabbinic Judaism, and why the liberal rabbis at Pittsburgh wanted to give nineteenth-century critical thinkers an ideological platform on which to stand. While those who want to understand God as personal are on solid historical ground, those who want to think of God more abstractly are, too.

Classic mysticism claims that the reality we experience is an illusion hiding the true nature of being within. The mystics, for all the naiveté of their elaborate symbolic systems, at heart were right: all is one. Look at an aerial photograph of a great river snaking its way through the countryside, leaving traces of where its banks used to be like scars on the landscape. Does the river shape its banks, or do the banks define the river? Yes. Both. Neither makes sense without the other. It is one system. So, too, with the physicists' and other scientists' description of how the universe evolved from an initial big bang (itself a metaphor, for there was no sound) to differentiated elements, galaxies and planets, and eventually to our earth, with its oceans and mountains, one-celled animals and fish that crawled out on land to become, eons later, you and me. Matter translates into energy and energy back into matter, Einstein demonstrated. Nothing is ever lost. It is all one massive system, from subatomic level to enormous galaxies millions of light-years apart. Does the order, including the laws of physics, shape what is, or does existence define the laws of nature? Yes. Both. No world, no element, no creature makes sense apart from the laws of nature, and neither do we know those laws except as inherent in being. God can be understood as this Order (as opposed to a metaphysical entity who created the order).

But what of *dis*order? the student of physics—or theology—might ask. Einstein went to his grave convinced that there must be some larger unified field theory that could bring it all into one equation (yes! the conditioned Jewish response in me affirms: *Shema Yisrael ... Adonai echad*, "Hear, O Israel ... God is one"[5]). Whether a single principle of organization accounts for all being, or several interrelating principles, it is one *system*. *Adonai echad*. God is one.

ARE PHILOSOPHICAL-GOD IDEAS ATHEISM IN DISGUISE?

But that is not true religion, a couple of noted thinkers, Daniel Dennett and Richard Dawkins, philosopher of science and scientist, insist in recent books that aim to debunk religion. Dennett argues that when the problems in maintaining belief in the personal God mount (or what I would call the problems in maintaining that personal-God metaphor), some of us, refusing to admit we are atheists, fall back on philosophical-God language. We do not really believe in God, Dennett accuses; we believe in belief![6] Dawkins quotes Dennett approvingly, admitting, though, that if being religious simply means assenting to Albert Einstein's affirmation that, "To sense that behind anything that can be experienced there is something that our mind cannot grasp and whose beauty and sublimity reaches us only indirectly and as a feeble reflection, this is religiousness," then he, too, is religious, adding only the caveat that what our minds cannot grasp now they may grasp in the future. Still, the skeptics insist, this is not "the God of the Bible, of priests, mullahs and rabbis, and of ordinary language."[7] Yes, it is! Or at least for many of us it is. How can Dennett write off Philo, Maimonides, and Spinoza, modern process theologians, and many others of us who argue for philosophical-God concepts? Moreover, he apologizes for being unable to deal with Eastern religions, which can be less theistic.[8] Even where Judaism, Christianity, and Islam are concerned, he ignores the mystical unity traditions such as Kabbalah. Both the personal-God theologians and the philosophical-God theologians admit that within the limits of our finite minds and human language we cannot fully or accurately describe God. Maimonides went so far as to insist that the very act of defining presumes to limit the infinite God, and thus we can only say what God is not. Unable to say what God is, we resort to the approximations of figurative, metaphorical speech.

As a rabbi, I claim as much right to define authentic religion as the self-proclaimed atheists. There is nothing inherently foolish or naive about religion. It is a paradigm, a mental framework, that can help make sense of a potentially overwhelming world of experience and choices. Scientists, some of whom are religious, have their paradigms, too. None of us, with finite minds, can compass the infinite possibilities. As Shakespeare put it, "There are more things in heaven and earth, Horatio, than are dreamt of in your philosophy" (*Hamlet*, act 1, scene 5).

Some of the faithful, feeling threatened, perhaps, by the mere suggestion that the challenges of theodicy and of science have weakened personal-God theism, will object, "But I don't want a philosophical God! I want a God who knows and loves me." The easy answer is: you are welcome to that God concept; we humbly agree neither of us fully understands God. The Rabbis of the midrash report that their students, hearing easy answers to difficult questions, would say, "You put him off with a straw. How would you answer us?" To the student yearning for a more emotionally involved deity, I would still claim lack of ultimate understanding but would add that because we are not in charge in the universe, we get the God who is, not the God we wish for.

I find the quest for God despite our limitations poetically expressed in the wonderful story in chapter 33 of Exodus. Moses, after the golden calf incident in chapter 32 and so many other exasperating incidents as he led the people, seeks reassurance from God. God promises to remain with him and lighten his burden. Moses remains unsatisfied. He wants to truly see, and thus fully understand, God:

> "Oh, let me behold Your Presence!" And He answered, "I will make all My goodness pass before you, and I will proclaim before you the name Lord, and the grace that I grant and the compassion that I show. But," He said, "you cannot see My face, for man may not see Me and live." And the Lord said, "See, there is a place near Me. Station yourself on the rock and, as My Presence passes by, I will put you in a cleft of the rock and shield you with My hand until I have passed by. Then I will take my

hand away and you will see My back; but My face must
not be seen."

(EXODUS 33:18–23)

We do not believe God has a hand, back, or face. This is
metaphorical language, myth. It means that with our finite
understanding we cannot fully grasp the infinite God, the God
large enough to be the one God of all that is. So we resort
to metaphor, often to the human image, as in this passage
and so many others—God as Creator, King, Father, Judge.
Not infrequently, though, and this is another indication that
it is all metaphor, the Hebrew Bible employs nonhuman
images—God as rock, eagle, or ever-flowing fountain, for
example.[9] Multiplying metaphors is an implicit admission that
no one metaphor is adequate. But of the God who or that is
vastly beyond our finite minds and metaphors we can only
catch glimpses. Each glimpse, and each metaphor, is partial,
an impression of our experience and yearning, not an adequate
portrait of God.

As I suggested in the previous chapter, people are going
to find a framework, or paradigm, to guide their thinking. I
believe our religious traditions, with their spiritual and ethical
guidance, are worth preserving. There is infinite value (pun
intended) in their scriptures, and wonderful aesthetic delight to
be had from all the arts that they have inspired and continue
to inspire. Furthermore, in mass society more than ever, the
communities and virtual communities that gather round these
traditions are also useful. Religious communities are not
perfect, but secular alternatives are scarcely more reliable.

GOD AND THE INTENTIONAL STANCE

But why—a question I struggled with for years—is the personal
metaphor so much more common when even back in Job and
Ecclesiastes there were those who saw its flaws? In part, no

doubt, because impersonal metaphors are just that: impersonal. A loving father is a more appealing image than a rock, an eagle, or a fountain, which refer to the same God but do not connote something that cares personally for *me*. Still, one who cares personally for me would not give me or, worse, my children incurable diseases or allow earthquakes and tsunamis to sweep away the innocent. We need the larger paradigms to make sense of and find meaning in our lives, yet need not cling exclusively to theologies and metaphors for God, which presume an unbelievably personal God, often the God of ever-diminishing gaps in our understanding.

One day, while I was pondering cognitive studies, it occurred to me why so many people insist not only that they react to God, but that God personally—individually, not as a force would but as a person would—relates to them.

The brain's main task is to keep us going safely through life. It monitors internal states, telling us, for example, when we need food or rest. It analyzes sensory input and reasons from past experience to enable us literally to navigate to where (carrying on the same examples) the food or rest should be available. To take better advantage of opportunities to satisfy our basic needs and, no less vitally, to avoid dangers (don't get eaten by that lion; don't let that suspicious-looking guy rob you), we further developed what some in cognitive studies call "theory of mind," or TOM. By analogy with our own minds, we try to figure out what others are thinking. TOM "is the ability to observe behavior and then infer the unobservable mental state that is causing it."[10] You take your date out to dinner, and as you linger over coffee, you get one message if she keeps looking soulfully into your eyes and a different message if she keeps glancing at her watch. You meet your new neighbor and find he has a dog. Does the dog wag his tail and then roll over to have his belly rubbed, or does it glare at you menacingly, tense up, and growl? In such cases you infer someone's intentions in order to respond appropriately. TOM is present in toddlers before age two, and it is fully developed by age four.[11] It is a basic survival skill.

As we gain more experience, our minds not only divine the intention of another toward us but also begin to imagine how our own alternative behaviors will influence the other's actions. Not only do I have an idea of your mental state, but I also know that you have one about mine. This can reach multiple levels of abstraction. If I move my checkers piece here, you will probably jump me there, enabling me to double jump you in return, except that I know you are a fine checkers player, so you will probably know that I am trying to trick you into a strategic error—and I also know that you know that I would not take you for a fool, so perhaps I am really trying to get you to do something entirely different. "Never let them see you sweat" is common advice. In a tense situation in business, romance, or really any social interaction, we manage our appearance, calculate the impact of every move before making it, and generally assume others are doing the same. Experiments on animals show that some get to a first- or second-degree level of abstraction (if I look like I'm about to fight, the other may back down), but probably only humans have the ability to keep track of four, five, and even six levels of intentionality.[12]

For my purposes here, the exact ways in which theory of mind can play itself out in complex levels of intentionality is not the point. The point, rather, is that normal brains do this at a basic level automatically. We constantly calculate what might happen to us, or has happened to us, and how we might react to better the potential outcome. For the brain, trying to get each of us safely and successfully through life, it is "all about me."

Evolutionary biologists explain that this habit of mind has survival value. A man walking through the forest hears a sound that stands out against the background noise of wind in the leaves and occasional birdcalls, say a snap that a twig might make. "A predator is stalking me!" he thinks and runs away. If it was a predator, he has saved his life. If not, he has wasted a little energy but done himself no serious harm. If, on the other hand, he said to himself, "No big deal; probably just a possum or something," he may be right, or he may be a lion's lunch. Three out of four of those outcomes are fine. The only dangerous option is thinking, "It's not about me." Those who take everything personally are apt to live longer and have more offspring.

So strong is the mental habit of personalizing, seeing intentionality directed at us in everything around us, that we even extend it to inanimate objects and forces. "Is the person walking toward me on a dark street a mugger?" is one thing. That is just the "lion's lunch" scenario in an urban setting. Consider some very different examples, though. The handyman installing a new fan in my house kept talking to the parts he was assembling: "Now get in there; don't fight me!" I would never have installed that fan myself, incidentally. The handyman has a feel for things mechanical, whereas I am convinced that machines don't like me. When the weather is unpredictable, my wife and I are apt to say, "If we bring the umbrellas, it won't rain!" Do we really think we can outsmart Mother Nature as we might a chess opponent? Coil a length of wire into a spring and press it down. Does it push back? Does it "want" to return to its original form? We take what Daniel Dennett calls the "intentional stance" not only toward things that actually do have intentions, but also toward things that clearly do not—though sometimes they may be more understandable for most of us if explained that way. "The basic strategy of the intentional stance is to treat the entity in question as an agent, in order to predict—and thereby explain, in one sense—its actions or moves."[13] Of course I know, intellectually, as does my handyman, that neither machines nor the weather have any intentions or feelings regarding us. But I do recall being caught unprepared by many a thunderstorm. Why take chances? Our brains are emotional as well as rational and take their jobs of getting us safely and successfully through life seriously. So we constantly personalize, thinking of things, situations, and natural forces as if they were consciously directed at us.

Taking the intentional stance, this universal human habit of mind, is more often than not pragmatically useful and has been written about at length.[14] For our purposes here, however, more detail will only be a distraction from the main point. We see the grandeur of nature and think it is about us, directed at us, a gift to us. We study hard and ace the test, or discover gold in the backyard, and jump to the conclusion that "Somebody up there likes me." There is no proving or disproving that. The more likely explanation, however, is that the grade is primarily a result of our own effort and the gold discovery a matter of blind luck. We burn a finger on the hot stove (probably after being warned more than once to be careful of it) or are diagnosed with cancer. "God is punishing me," we think. Again, there can be no proving or disproving that. But the laws of physics and biology—yes, and blind chance—or our own foolishness (Are we careful? Do we smoke?) are more likely explanations.

We are sometimes beneficiaries and sometimes victims of the divine order. God as Order is real, not of our making, and has been shaping all that is since before we and our species came into existence. As such, God is ultimately related to learning, gold, burned fingers, and cancer, ultimately related to all that is, all that we regard as good, evil, or neutral. God as caring personality may be real, too. Each is one take on divinity, a concept, a metaphor, imperfect and inexact. As we ponder the implications for religion of the operations of our brains, moreover, we shall need to consider in due course the ethical implications of that Order and whether, for those more comfortable with a philosophical-God concept, there is any point in prayer. For now I am merely trying to answer the question above: why is the personal metaphor for God so much more common than impersonal and philosophical metaphors? The answer, which should by now be clear, is that our brains, for good reasons that contribute to our survival, have evolved to see intentionality even where there may be none, or, as I have phrased it more simply, we personalize.

3

Mystical and Spiritual, Neurological and Theological

What is religious experience as it plays out in our brains? We begin by examining mysticism and more conventional religion. Scientifically, we find, religious experience is real, a demonstrable alteration of consciousness. What we call "spirituality," moreover, is not just a warm, fuzzy feeling, but the result of emotion reaching a perceptible level as we filter experience through our religious paradigm.

> *The human spirit is the lamp of God,*
> *Revealing the deepest self.*
> Proverbs 20:27[1]

*A non-mystic is someone who believes that when truth
is explained to him in words, he should understand
that truth. The mystic is someone who believes
that real truth, meaningful truth, can never be fully
expressed in words.*

Joseph Dan[2]

As my wife and I sit down to dinner on Friday evenings, we begin the Jewish Sabbath with the traditional rituals. She lights the candles, praising God for "hallowing our lives with commandments," including this simple ritual. I sing the *Kiddush*, the blessing over wine. Together we sing the blessing over the challah, the special loaf of braided Sabbath bread, and then taste it. Through the alchemy of ritual, more than rational thought is engaged, though if we think about it any particular week we do know that "light is the symbol of the divine,"[3] that wine cheers the human heart (Psalm 104:15), and that challah deliciously connects us with the ancient sacrificial worship of the Torah (Numbers 15:17–21). Candle flames entrance people in virtually every religious tradition, suggesting, as here, the presence of God, or elsewhere recalling ancient miracles or commemorating ancestors. Whatever the neurological basis, and there must be one, the candles especially and the whole ritual effectively set a mood of holiness (*kedushah*, from the Hebrew root *kadosh*, something set aside as special), setting this meal apart from other meals. In some families, children are blessed by the parents at this time, and spouses sing one another's praises. As often as not we appreciate the symbolism without pausing to reflect on it. It comes up every week, after all, and we have to get to worship services in a little while. But if we were to omit this ceremony some Friday evening, we would miss it. Engaging our minds sometimes more deeply and sometimes less, the ritual employs all our senses to create what contemporary American culture calls a spiritual moment.

Traditional rituals provide religious experience. I like to think of ritual as the flesh on the bare bones of theology. Most obviously, people join in public worship, where prayers, scripture readings, and often sermons, not to mention choreography, silent meditation, and music of all varieties, occupy us. All this, moreover, is commonly reinforced by special architecture and visual symbols and frequently by the special dress of officiants, participants, or both. Depending in part on the skill of the "choreographers" and leaders and in larger measure on the devotion of the participants, and even on their mood that particular day, for the various worshipers the exercise may be more or less effective, and even sometimes a total failure, in evoking spirituality.

Many other experiences are regarded as worship. Many study, especially sacred texts, together. Religious communities work together in idealistic social projects. Even social activities that may lack any direct reference to God still serve to create community, thereby reinforcing religious identity and a sense of belonging. We can call all that we do individually and collectively in the name of religion "religious experience." But most of us would not call all of it, and certainly not modern synagogues' or churches' endless round of committee and board meetings, "spiritual." Spirituality, it seems, has more to do with God or at least with some sort of self-transcendence.

Recognizing such obvious religious experiences has pragmatic value but misses the radical challenge of most religions, and certainly of Jewish-Christian-Islamic-style Western religion. The distinction between religious and secular spheres of life and society is modern and by traditional religious thought is a false one. While Judaism traditionally counts 613 commandments, we can look more conveniently at the Holiness Code in Leviticus 19, which the Rabbis regarded as containing the essentials of Torah.[4] There we find the text going back and forth indiscriminately between ritual and ethical commandments. It begins with honoring parents, immediately turns to eschewing idolatry and properly offering sacrifices (the prime form of biblical worship), then goes on to a host of ethical injunctions, among them caring for the poor and the stranger, paying laborers on time, and loving one's neighbor as oneself, turning later to farming properly, going to the priests to make atonement when these commandments are violated, and so on and so forth. All these things and more—all of life—is, by the biblical tradition, the province of religion. So if religious experience is what is done as part of religious living, *everything* is religious. Disestablishing religion in many modern nations created a de facto distinction between religious and secular, and private and public. This made religious pluralism more workable. But all the religions continue to preach that their ideals and values should shape the behavior of their adherents not only at home and in houses of worship but also in the workplace, the political realm, and so on—everywhere, not only where specifically ritual activities occur.

To complicate matters further, most of us know people who claim to be "spiritual, but not religious." One recent poll found that in America, "the old terms—'atheist' and 'agnostic'—are no longer catch-alls for everyone outside traditional belief. In fact, 24% of respondents put themselves into a whole new category: 'spiritual but not religious.'"[5] Then there are subcategories among religious believers who see themselves as more rationally inclined or more mystically inclined. There is great variety even within each major religious tradition, as well as between them. "Religious experience" may reasonably be defined as that which is done in the name of God or religion, but such a definition tells us little. Furthermore, as many have lamented for years, "spirituality" is such a fuzzy term that few are quite sure what it means. We tend to regard spirituality as the sense of being in the presence of God, though that may leave out those who claim to be spiritual without being religious. Even as we remain vague about the meaning of the term, a chorus of voices announces a growing modern "hunger for spirituality."

How nice it would be if the cognitive studies researchers could tell us what is going on in our brains when we say we are having a religious experience. This might require the theologically hazardous step of ignoring the classic religious message that religion is for the whole of life. Is someone who is, say, helping the proverbial old lady across the street having a different experience if he thinks of that as a religious obligation, a way to serve God, as opposed to simply a kind deed? I suspect not, but more data is needed to back up that hunch. Could we not at least determine whether different regions of the brain are in use, or in use with a different balance among them, when we think we are doing something specifically religious, prayer or other ritual or Bible study, as opposed to participating in the civic ritual of a Fourth of July parade or studying for a final exam in grammar? I suspect that the balance among brain processes is altered, but, again, this remains speculative. Such studies have begun, however. Together with other aspects of cognitive studies, I suggest they can be helpful in moving us toward a less fuzzy understanding of spirituality.

DO WE HAVE DIRECT EXPERIENCE OF GOD?

Millions of people have a sense of relating to God, and some even of experiencing God's presence. One cannot touch, see, smell, hear, or taste that which is nonphysical. We do, though, experience some of our own thoughts, which are physical, but only insofar as electrical patterns playing on brain circuits are physical. How we might sense God's presence is difficult to say. One logical possibility is that God—as pure spirit or pure intelligence—plants ideas directly into our brains, bypassing the senses. What is more likely, however, and does not require such a supernatural explanation, is that we have certain experiences in the natural course of living that give us that subjective sense. The example I used at the beginning of chapter 2 was awe in nature. At the beginning of this chapter I spoke of the difficult-to-express but very commonplace reaction, probably both rational and emotional, to seeing candles burn and tasting, smelling, and hearing other physical things, such as bread, wine, and blessings, which we then, like any other experience, filter through our higher-level mental paradigms (discussed at the end of chapter 1). Let us remain for a minute with the explicitly religious example. My wife and I do not get our sense of holiness on the eve of the Sabbath exclusively from the physical acts—the saying or singing of prayers or the seeing, smelling, and tasting of symbols. We have both experiential knowledge (I loved this as a child) and abstract understanding ("light is the symbol of the Divine") of Sabbath rituals. We could not now have the experience, even if we wanted to, without it activating memory circuits in our brains that are also connected with emotional circuits. So we do not need a *deus ex machina*, a divine intervention, to explain why doing Jewish things makes us feel Jewish, even on occasion to the degree that we might say, "What a spiritual moment!" as if we were directly experiencing God.

These religious reactions, from the pedestrian but important, "Isn't this a nice, loving, family moment," to the rarer sense of being in God's presence, are conditioned reactions involving both abstract thoughts and emotions. (As we shall see, moreover, thought and emotion are not the distinct categories people often assume them to be.) So by parents and teachers, and then as parents and teachers, we are conditioned to experience these symbols as religious—and not just pertaining to any religion, obviously, but to ours. Others react to different symbols—or in the case of candles to the same symbol—similarly, albeit through the prism of their own religious/interpretive paradigm. In other words, when we say something is a special religious experience, even to the point, on rare occasions, of thinking "God is with us!" we are almost certainly doing so as a conditioned response. I say this not to trivialize something precious to us, but to point to its neurological underpinnings. The omnipresent God, of course, is present. Past conditioning and present circumstance have come together to heighten our awareness of that at the particular moment.

Because my religious tradition has taught me to do so, I can relate to the ritual itself as sacred because it points to God. Scripture, likewise, is sacred because it has taught so many generations, and me, too, about God. The sense of *kedushah*, holiness, does not always or even usually require the conviction that one is attuned, directly, to God. Sometimes it does; in my tradition more often it does not. The primary experience is of physical things (sunset, wine, prayer, story); the religious content is inferred. An inferred God or other inferred sense of holiness is a secondary manifestation of primary experience. The brain conditioned to do so interprets certain primary experiences as religious.

For some in each tradition, though, the desire for direct experience of God persists.[6] So mystics, over the ages, have become convinced that in arcane ways they can directly, as first-order experience, sense God, and thus truth and ultimate reality. In some cases they believe they somehow *merge with* God. There are a great variety of such theories or understandings, loosely lumped together by modern scholars as mysticism.

I would once have said that if you insist that reality as we perceive it is an illusion, hiding a deeper reality within, then you are a mystic. Then I started reading about physics, which empirically concludes precisely that what we perceive as solid (or liquid or gas) is made up of molecules, which consist of atoms, which in turn consist of subatomic particles, and includes more space than anything else. The same could be said of other branches of science, including neuroscience (what we experience as unified self is actually many processes going on simultaneously, and more is going on in the brain at any moment than reaches the threshold of conscious awareness). Scientists, though, think they communicate objective facts in words and mathematics. Mystics insist their experience of God is ineffable. So I have come to like Joseph Dan's definition in the epigraph of this chapter. The mystic believes that ultimate truth cannot be expressed. There may well be a touch of the mystical in all religion, as when I and others argue that God is so far beyond us that we can only speak of God metaphorically. Yet I believe the metaphors are genuinely helpful, that we can express something meaningful about God, stopping short of full mysticism. The point, to reiterate, is that many mystics insist that God *can* be experienced, though people lack the ability to communicate the experience to others.

This is an important digression because for many mystics the goal and highest achievement is not only to experience God, but also to experience a merger with God, a *unio mystica*, "union with God," and neurological research has shed light on what appears to be the neurological substrate of that phenomenon. The chapter on mysticism in William James's classic *The Varieties of Religious Experience* presents passage after passage in which mystics from various traditions speak of the effort to tame and even temporarily eliminate the mystic's private self or ego, ultimately reaching a sense of oneness with God or all being (and of course in pantheistic systems, which mystical theologies often are, everything is God, so union with being *is* union with God). James notes that mystics typically describe their experience and what they learn from it as ineffable.[7] The history of Jewish mysticism shows more complexity. Some mystics—Abulafia, Tzefat kabbalists, and to some extent modern Hasidism—seek union with God, but others, notably the medieval Spanish kabbalists, do not.[8] This is still mysticism, for even kabbalists not seeking a state in which the boundaries of the self are lost in the vast divine sea of being, the feeling often described as "oceanic,"[9] agree that ultimate reality may be glimpsed but never communicated.[10]

SCIENTISTS STUDY MYSTICAL EXPERIENCE

Doctors Andrew Newberg and Eugene d'Aquili of the University of Pennsylvania received permission to study Buddhist monks as they meditated, attempting to reach a mystical state. They tell us of the meditation of a monk named Robert:

> First, he says, his conscious mind quiets, allowing a deeper, simpler part of himself to emerge. Robert believes that this inner self is the truest part of who he is, the part that never changes. For Robert, this inner self is not a metaphor or an attitude; it is literal, constant, and real. It is what remains when worries, fears, desires, and all other preoccupations of the conscious mind are stripped away. He considers this inner self the very essence of his being. If pressed, he might even call it his soul.
>
> Whatever Robert calls this deeper consciousness, he claims that when it emerges during those moments of meditation when he is most completely absorbed in looking inward, he suddenly understands that his inner self is not an isolated entity, but that he is inextricably connected to all of creation. Yet when he tries to put this intensely personal insight into words, he finds himself falling back on familiar clichés that have been employed for centuries to express the elusive nature of spiritual experience. "There's a sense of timelessness and infinity," he might say. "It feels like I am part of everyone and everything in existence."[11]

Newberg, a radiologist as well as a religion professor, used nuclear medicine to take pictures of the monks' brains as they reached what they reported as their deepest meditative levels. A radioactive dye was injected into the subjects' bloodstream, enabling imaging of where blood went in greater or lesser than usual amounts in the brain, indicating greater or lesser use of certain areas. Each of us has what Newberg calls (along with the technical neurological terminology) "a small lump of gray matter nestled in the top rear section of the brain" that orients us in physical space. It keeps track of where we are, our boundaries, so that as we move around we do not bump into things and we appreciate how near or far other things are. "In simple terms, it must draw a sharp distinction between the individual and everything else, to sort out the you from the infinite not-you that makes up the rest of the universe."[12] As Robert went deeper and deeper into his meditative trance, there was less and less activity in this area. The doctors speculate that it was not that the brain area in question was turned off, but rather that it was receiving less data than usual from the senses. Lacking data that showed where the monk ended and the world began, his brain lost track of that distinction. He felt like an undifferentiated part of the vast sea of being.

These results were replicated first with other monks meditating in Tibetan Buddhist style. Then Newberg and d'Aquili repeated the experiment with Franciscan nuns involved in meditative "centering prayers." "The nun begins by focusing her mind on a particular prayer, word, or passage from the Bible. Then she closes her eyes and reflects on the inner meaning and spirit of the text. As [one sister] explained, 'I open myself to God's presence.'"[13] Because the nuns concentrated on the words of their prayers, their language centers were activated more than those of the monks. But they, too, had a marked decrease in the orientation center, the intensity and length of the prayer experience inducing them to lose track of space and time and seemingly to lose individuality, which was in turn interpreted as being directly in contact with God.[14] In other words, the mystical experience of *unio mystica* was very similar, though the brains of the Buddhists interpreted that experience as joining a universal consciousness and the brains of the Christians interpreted it as communion with God.

In a more recent book, a more secular scientist, Harvard neuroanatomist Jill Bolte Taylor, recounts her experience having a stroke.[15] A stroke is caused by the interruption of blood flow to some part of the brain. Dr. Taylor, it was later determined, had a congenital problem with blood vessels in the left side of her brain, where, among many other things, the orientation area that became less active in the meditating brains studied by Newberg and d'Aquili is located. In Taylor's stroke, instead of receiving less data from the senses as it continued to function, that orientation function itself was impaired, as was much more. The left side of the brain generally handles linguistic, mathematical, and analytic functions and keeps track of narrative time, whereas, Dr. Taylor explains, the right side creates a "master collage of what this moment in time looks like, sounds like, tastes like, smells like, and feels like."[16] The two sides typically work so well together that we cannot introspectively distinguish between them; we feel like one whole person, not two personalities in one brain. Still, some of us (including, quite obviously, this theological writer) are more left-brained and analytical, and some of us (most artists) are more right-brained and spontaneous. Dr. Taylor is convinced that as she had her stroke and then struggled long term to overcome the disabilities it caused, more was involved in her altered state of consciousness than just the orientation function that Newberg and d'Aquili emphasize. In particular, impairment of her language and time-keeping abilities were also involved.[17] The important point for the moment, though, is that her experience parallels the mystical experiences we have been discussing:

I remember that first day of my stroke with terrific bitter-sweetness. In the absence of the normal functioning of my left orientation area, my perception of my physical boundaries was no longer limited to where my skin met air. I felt like a genie liberated from its bottle. The energy of my spirit seemed to flow like a great whale gliding through a sea of silent euphoria. Finer than the finest of pleasures we can experience as physical beings, this absence of physical boundaries was one of glorious bliss....

Without a language center telling me: "I am Dr. Jill Bolte Taylor. I am a neuroanatomist. I live at this address and can be reached at this phone number," I felt no obligation to being her anymore. It was truly a bizarre shift in perception....

When I lost my left hemisphere and its language centers, I also lost the clock that would break my moments into consecutive brief instances. Instead of having my moments prematurely stunted, they became open-ended, and I felt no rush to do anything.... I morphed from feeling small and isolated to feeling enormous and expansive....[18]

This and many other passages in *My Stroke of Insight*, Dr. Taylor's enthralling book, sound more like contemporary American "new age spirituality" than classic mysticism. She uses terms such as, here, "the energy of my spirit" and elsewhere "bliss" and even "Nirvana," but not "God." There is little doubt, though, that the phenomenon is the same, albeit interpreted through a different higher-level brain paradigm. For these three examples, Buddhist, Christian, and secular, one could go round and round in chicken-and-egg fashion trying to say whether the experience is primary and the interpretation secondary or whether the interpretation the brain adds determines what individuals experience or at least what they express and remember. Either way, we—our brains—assign the meaning. The paradigm we employ to interpret helps shape the experience as well as our understanding of it.

This is important because it explains a widely reported religious phenomenon neurologically, demonstrating that the same brain mechanisms are involved in different religious—and not so religious—traditions. That further demonstrates that the same primary experience takes on different meanings based on the individual's higher-level paradigm or filter. Furthermore, we do not have to assume the nature of God, or even the existence of God, to appreciate that something real is going on in the brain and consciousness of those who report that they have had this "oceanic" experience of oneness with all. The findings, however, do cast into doubt explanations of *unio mystica* as out-of-body experiences, as the soul, for instance, rises to higher and higher levels of heaven. (Mystics would probably reply that such language is metaphorical in the first place; the mystic experience—once again—cannot be communicated with any precision.) What remains an open question, however, is of the essence theologically. Is *unio mystica* an illusion, a trick the brain plays on believers? Or, given the scientific understanding of all matter and energy having started in one big bang (or even perhaps a series of them as the universe expanded and contracted), are the mystics overcoming an opposite illusion, namely that they are distinct entities when in fact they are, we all are, part of one giant Order? The stroke-stricken scientist joins classic mystics arguing precisely that:

> Of course I am fluid! Everything around us, about us, among us, within us, and between us is made up of atoms and molecules vibrating in space. Although the ego center of our language center prefers defining our *self* as individual and solid, most of us are aware that we are made up of trillions of cells, gallons of water, and ultimately everything about us exists in a constant and dynamic state of activity.[19]

Both are logical, arguable explanations. It depends on your paradigm for understanding life and the world. That the universe is one vast sea of evolving being subsuming us is basic physics. Within it, though, individual entities come and go, and they, too, from galaxies to microbes, and us in between, are real. A fortunate few have consciousness, which at the human level privileges us with the opportunity to understand much, but still not all, of the rest. That is real, not an illusion. Cosmic oneness, the universe united by the laws of nature, is also real. Because I mean that literally, and the definition of a mystic is that he or she does not believe "that real truth, meaningful truth, can be conveyed in words," I say that without pretending to be a mystic. Reality—thank God—is as wondrous as anything the generations have dreamed.

Do we, then, have direct experience of God? Suddenly we are back to our understanding of that which is beyond our full understanding. Do we experience a Personality who "knows" as we know, "creates" as we create, "rules" as a king or president rules? I have yet to see convincing proof or disproof. If by God, however, we mean the cosmic Order of which we are a part, of course we experience it, every day. Our whole existence depends on it—or, if you will, on It.

EMOTION AND SPIRITUALITY

We should not lose sight of the fact that the mystics, and even the stroke victim, claim as vivid personal experience what I just treated as abstract concepts to be understood intellectually. The example of my wife's and my Sabbath experience with which I began, moreover, and the analogous ritual experience millions of believers have had in all sorts of ritual religious contexts, makes no such extravagant claims of divine intoxication (or, at least, rarely does so). Neither are such examples the same as but merely to a lesser extreme the experiences of the monks and nuns in the study, who devote large chunks of time—after years of practice—to achieve what they believe is a spiritual breakthrough to another level of reality. Even with sufficient time, meditators do not always succeed, and when they do, they insist the experience is incommunicable. Much of what most of us more commonly claim as religious experience—prayer and other ritual, text study, the doing of good deeds, awe in nature—may often be difficult but still not impossible to communicate. And though the religious feeling in these activities may not always happen when we wish, it sometimes happens when we least expect it. Many a poet or preacher communicates such religious experience quite well, moreover, at least to those prepared to hear it. I suggest that as mystical experience has a neurological substrate, our more common and pedestrian religious experiences must as well. So I hope I have chipped away at the question of what religious experience and spirituality are. But we must press on. What is happening, neurologically, in the more common religious experience so many of us believe we have?

We *Homo sapiens* are distinguished by the sapient, cognitive, ruminative, in a word *rational* part of ourselves. We know that we evolved from the animal kingdom and, as embodied creatures, have an animal nature. But the part of our brains where we do our abstracting, roughly speaking the frontal cortex behind and around the forehead, is larger and more developed in people than in any other animal and is particularly rich in neural connections to other areas so that decisions can be made, in effect, as part of a network. By contrast with this rational part of us, we have tended to think of our emotions as lower and in need of control. Dr. Antonio Damasio of the University of Southern California, though, explains that although emotions mostly originate in other parts of the brain, they make vital contributions to our rational thinking. They are in the network—part of, not other than, our thinking. You meet a lot of people in the course of a day and tend to forget most of them. But you will certainly not forget the one you found especially sexy or generous or whose conversation either particularly fascinated you or thoroughly aggravated you. When an emotion is triggered, a memory is, as it were, flagged as important for your future reference.

We depend on emotional input to thought in other ways, too. You are walking along when suddenly you see someone or something rushing at you—a raging elephant, a speeding Chevrolet, a menacing character coming out of the alley. This is not an ordinary experience, and your mind is seized by a very helpful emotional reaction: fight or flight. This is a matter of survival, and there is no time for your big brain to calmly reflect; you must act! We tend to like some emotions—love, awe, and joy, for instance—and might prefer to avoid others, such as fear, anger, and sadness. These and other emotions play a direct role in motivating our attention and proper—in the sense of helpful for our well-being—reaction. We would not want to be blasé when endangered, downcast when the home team wins, or exultant when a loved one dies. Emotions infuse and often direct appropriate rational thinking and physical action. As a matter of fact, Damasio demonstrates, our emotions consistently influence us physically; they are "embodied." When you react to a threat, blood flows to the muscles and adrenaline pumps so you are ready for action. As readers of body language, we instantly distinguish friendly, fearful, and angry faces and postures. If you have autism and thus do not, you will likely have social adjustment problems. When we speak of fear as a feeling in our guts and of behaving courageously as "having guts," this is by no means purely metaphor. Fear and grief do manifest themselves viscerally ("in the gut"). And without the physical reaction to sexual attraction, the race would not continue. On a more subtle level, recall the Capgras syndrome patients in chapter 1 who recognize parents not only by appearance or voice but also by their own unconscious emotional reaction measurable as galvanic skin response—a bit of sweat. Damasio engagingly speaks of "the motion in emotion." We react both physically and intellectually to our emotions. "Are you having a good day today?"—or a bad one—is less a matter of rational judgment than of emotional reaction to what has happened to you. Because the emotions come from older and deeper places in the brain (in terms of evolution), they amount to a parallel system of thought that influences the more abstract and logical system of thought we call "rational." To some extent we learn to recognize rationally and partially control (inhibiting or magnifying) our emotional reactions, but that extent is limited; we always operate on both levels at once. We keep thinking all the time but experience our emotions as waxing and waning in response to what we are experiencing (externally) or thinking (internally) at any given moment. Modern neuroscience is not only mapping the several sources of emotional reaction in the brain and how they trigger our reactions, but it is also trying to explain how we, as humans and thus more sophisticated than lower animals, recognize these states.[20]

What has all this to do with religious experience and spirituality? Everything, I suggest. When we occupy ourselves with specifically religious occupations, however different religious traditions and individuals may define that (most offer prayers, do loving deeds, study inherited texts, while only some offer sacrifices or sit for hours in silence), we call that, in the simplest sense, religious experience. To a greater or lesser extent we may intellectualize about it, noting special procedures and symbolic meanings. This is also religious experience. But at some point we sense something more, a shift in subjective feeling, a breakthrough, *spirituality*. At that point some would say, and some would not, that we are sensing God. But our question here is less highfalutin. What is happening neurologically? Emotion is always there, at least in the background, but I suggest that when a surge in positive emotion is triggered by an experience, to the extent that it breaks through powerfully enough to become conscious, we experience the moment as different, special, set aside—the root meaning of the Hebrew *kadosh*, "holy." This experience associated with religion comes to be felt as *spiritual*. We become excited or awed, feel connected to community, tradition, or even God. Another term for this is self-transcendence. Obviously, I am not literally connected to the text as, say, the story of Sinai or Isaiah's throne vision transfixes, literally excites, my attention. I am not literally connected to my religious community as we sing a powerful hymn together, but I have a sense of enhanced bonding, a quasi-mystical feeling of being part of something larger than myself. We do not usually mean it literally when we speak of being "inspired," either, though occasionally we may suddenly gasp—inhale—from our sense of wonder or excitement over something. The term "inspiration" (in-spiration; breathing in) derives from that emotional response. This is all, in effect, "the motion in emotion" in a religious context. It is not purely emotional, for the ritual, prayer, or study takes place in a learned and rationally appreciated context, one's religious tradition, else it is not recognized as religious. Emotion *and* rationality are involved.

WHEN EMOTION REACHES CONSCIOUSNESS— IN A RELIGIOUS CONTEXT

The issue, it seems to me, is the balance between emotion and rationality. Neuroscientists are beginning to explore this experimentally. A study of mystical experience found enhanced activity in emotional brain centers (not surprising as one "dials down" analytic thought to feel the moment),[21] and a study of six Evangelical Christians and six secularists reading the Twenty-Third Psalm, a nursery rhyme, and the phone book (with six more secularists as a control group) found the religious subjects had enhanced activity in rational brain centers with the psalm. (Researchers hypothesized that the religious individuals were placing the familiar text into their established religious "schema"—what I have been calling their paradigm—a more rational function.)[22] My hypothesis about more conventional religious experience echoes Newberg's about meditative practices.[23] Both rational and emotional areas of the brain are involved, but the powerful *feelings* (note the word) of holiness, spirituality, and transcendence occur when more emotion than usual is stimulated. Psychologists Malcolm Jeeves and Warren S. Brown have summarized the research similarly.[24] Research continues. My hunch is that experiments will show that "Isn't that interesting?" becomes "Wow, how profound, how meaningful!" precisely when emotional arousal becomes conscious.

Spirituality, then, whatever the balance turns out to be as it is further studied, is always emotional as well as rational. We *feel* a bond with community, tradition, or even God and may think we thereby transcend personal limitations, no longer feeling as small or alone. Spirituality is subjective. Spirituality that is not felt, not subjective, is not spirituality at all. Furthermore, because "all emotions use the body as their theater,"[25] and spirituality has a major emotional as well as a rational component, spirituality is not exclusively in the mind. As emotion has a physical component, spirituality does, too. There is often more than metaphor involved when we speak of being "moved" or "deeply touched" by a religious experience.

Whether felt spirituality is a response to something objectively real is not in doubt where community and ancient texts, as well as ritual items, sacred music, and other aspects of our traditions, are concerned. My candles and blessings on Shabbat are very real. They suggest and symbolize, and I respond. But what of that which is symbolized? The Sabbath is real. Tradition is real. But is God, for those who say they sense God's presence, literally, physically, *objectively* and not only subjectively felt? That is unanswerable until we define what we understand God to be, and once again, all God-talk is metaphorical and approximate. Those who think of God as a physical presence (or as all being) will have no difficulty believing, even insisting, that their subjective sense is of Something or Someone objectively real. Those of us who think of God in more abstract, philosophical terms will likely say, as I would, that our spiritual sense is an intellectual excitement, not a direct experience of God unmediated by thought and paradigm.

That the examples of spirituality in this chapter are all positive, not negative, experiences is not accidental. In the book of Psalms, those crying out in physical or psychological pain, far from feeling God is the source of their woe, feel bereft of God ("How long will You hide Your face from me?" for instance, in Psalm 13:2, and many similar passages). In most religious paradigms, I suspect, we associate God more with good than with evil. Still, for monotheists everything comes from God, not only the good, so the theoretical possibility of negative spirituality remains. In a Hasidic tale, Rabbi Zusya complains that he loves God enough but does not fear God enough, and he requests that he might fear God as much as the angels do. "And God heard his prayer, and his name penetrated the hidden heart of Zusya as it does those of the angels. But Zusya crawled under the bed and howled like a little dog, and animal fear shook him until he howled, 'Let me be like Zusya again.' And God heard him this time also."[26] One wonders if worshipers hearing Jonathan Edwards preach "Sinners in the Hands of an Angry God," or others whose spirituality includes a major dose of fear, might similarly associate God with negative emotion. "Spirituality" seems too mild a term for terror, but I would not deny the potential for negative spirituality, though I suspect it is far less common than the positive variety.

This notion of spirituality as triggered by an emotional reaction to something, whether something external or even one's own thinking, is worth lingering over, for spirituality is an important aspect of religion. Several years ago I and a local minister led an interfaith group to Israel, and one of our stops was the Hill of the Beatitudes overlooking Yam Kinneret, better known to Christians as the Sea of Galilee. Here Jesus is reputed to have given his famous Sermon on the Mount (Matthew 5:1–7:29). There is a lovely church with beautifully manicured grounds and a fine view of the lake. Like so many other groups, we sat everyone down and the minister read the beatitudes, a portion of the sermon (Matthew 5:3–10). I had heard it before and found it, once again, a beautiful, idealistic message. For me, though, that was a largely intellectual response. I was not especially moved hearing the New Testament passage at that particular place. Recently I was talking to a ministerial colleague who had just led a group of ministers to Israel. They stopped at the same place, and her instructions from the tour operator included gathering the group together to read the same passage from Matthew. Her first reaction to the instructions was that the planners were trying to manipulate them; a group of ministers, surely, knew the passage and did not need to have it read to them. But she decided to read it anyway. And in that spot, with the Sea of Galilee below, she said, she could scarcely believe her own reaction to reading that text. "It was powerful!" she said with obvious affect.

Consider: this was not a purely emotional reaction, for without the rational understanding of the place and the text, the religious framework of the experience, it would not have been so powerful. But the knowledge was in her mind whether or not she had read the text. Reading the text triggered an emotional surge. On my trip to Israel I had a similar experience as our group arrived in Jerusalem and we stopped atop Mount Scopus to survey the city beneath. Taking a cue from a travel book, I had planned on reading ten lines of a famous poem in which the poet Avigdor Hameiri, in 1922, saluted the city from the same spot and, echoing the millennial yearning of Jews for Jerusalem, exclaimed, "For one hundred generations, have I dreamed of you," and concluded, "Jerusalem, Jerusalem, from your ruins, I will rebuild you."[27] I had been to the overlook before, and had read the poem before. But I had not read the poem there before, and as I was about to recite it, I realized I was too overcome with emotion and asked my Christian colleague to do so for me. It was not even, technically, a prayer, but in this context it was bound up in my religious identity. My colleague found it a meaningful reading (as I found the beatitudes a meaningful reading), but his text experienced in his religious paradigm, and my text experienced in mine, created a Christian spiritual moment for him and a Jewish spiritual moment for me. Spirituality kicks in with the often unconscious and usually irresistible message: pay attention; this is profound!

Similarly, when volunteers go to a homeless shelter to help serve a meal, a not very religiously inclined person, motivated simply to be a good and compassionate citizen, may feel good about the time spent and may well have some emotional reaction to the scene or to an individual's look or word of thanks, yet not identify that reaction as spiritual. Another person, doing the same thing because she regards it, if Jewish, as a *mitzvah* ("commandment," sacred duty) or, if Christian, as an act of love (for Jesus instructed that feeding the poor was tantamount to feeding him, in Matthew 25:35–40) may speak of having had a spiritual moment.

Consider a final example, this one of an activity not generally regarded as religious. An atheist friend and I are sitting at the symphony listening to an exceptionally fine performance of a brilliant violin concerto—Beethoven, Sibelius, pick your own favorite. We both love it. It was brilliantly composed and now sensitively performed to evoke emotional reaction. I say to my friend as the music ends and the applause begins to die down, "That was a religious experience!" And I mean it. My soul has been stirred. He responds (because he likes to needle me!), "I don't know what you mean by 'religious' in this context, but yes, it was glorious!" The point, once again, is that one's higher-level paradigm shapes both the experience and its meaning. It is probably a useful convention of common parlance to reserve not only the term "religious" but also the term "spiritual" for experiences that somehow call God or religion to mind. But recall what I earlier called the "radical challenge" of most religions. God is omnipresent, and religion, thus, is about the whole of life. My friend, though loath to employ the same language, had an experience very much like mine, though not absolutely identical because each was filtered through different mental paradigms. Given the similarity and given that lots of people these days claim to be "spiritual but not religious," it might be reasonable simply to say that whatever touches the human spirit is spiritual.

"Spirit" I would take, and so does my dictionary, as a synonym for "soul." But "soul" carries a lot more intellectual baggage in Western religion. So we turn next to the soul and immortality.

4

The Soul Which Thou Hast Given unto Me?

The Hebrew Bible's understanding of soul as life force was radically altered when Hebrew thought encountered Hellenistic mind-body dualism, enabling not only Christianity but also Rabbinic Judaism to begin to think of the soul as a metaphysical entity separable from the body. Modern science undermines that. We can make sense of much of our living and thinking, even of our yearnings for love, truth, and beauty, without recourse to the supernatural. "Soul" remains a powerful metaphor, but a metaphor cannot carry our consciousness beyond death. The biblical idea of immortality, corporate rather than individual, remains credible.

The soul which Thou, O God, hast given unto me came pure from Thee. Thou hast created it, Thou hast formed it, Thou hast breathed it into me; Thou hast preserved it in this body and, at the appointed time,

> *Thou wilt take it from this earth that it may enter*
> *upon life everlasting....*
>
> Union Prayer Book[1]

I grew up hearing that liturgical passage recited regularly in my synagogue. Do we have souls? We hear that a piece of music touches the soul, and for the dead we hear the pious wish "God rest his soul." Especially from the Christian side of our societal debate over abortion, there is concern over whether "ensoulment" takes place at conception or after some days or weeks. There is a rich folk and literary tradition of characters such as Marlowe's Dr. Faustus selling their souls to the devil for knowledge or other worldly advantage. Close friends and lovers speak of being "soul mates." As I was working on this chapter, a mailing from a Plano, Texas, senior facility arrived, featuring a photograph of two old men laughing together, with the caption "Friendships that touch the soul."

The liturgical passage presents the soul as a distinct entity created by God, added to the physical body created sexually by parents, and reclaimed at death. That idea of the soul will be familiar to Christians and Muslims, as well. Students of Jewish thought, though, will rightly protest that in biblical Hebrew thought, soul and body are inextricable. When, in biblical tales, life somehow seems to be restored after death, the whole person, body and soul, continues. Elisha revives the Shunamite woman's son in 2 Kings 4. When, in 1 Samuel 28, the necromancer at Ein Dor brings Samuel back from the dead to speak to Saul, he looks not like the ghosts of modern movies, but like "an old man ... wrapped in a robe" (verse 14). At the end of Elijah's life, his whole body, not a disembodied soul, is transported to heaven (2 Kings 2). These are rare incidents, moreover. Harvard biblical scholar Jon Levenson sums up the biblical position on the reality of each individual's death as "nicely and exhaustively stated by the woman of Tekoah to King David, 'We must all die; we are like water that is poured out on the ground and cannot be gathered up' (2 Samuel 14:14)."[2]

With a significant exception (of which more in a moment), classic Rabbinic thought carried on the inextricability of body and soul. Orthodox Judaism to this day teaches that when the Messiah comes, the dead will be physically resurrected. My Reform Jewish forebears viewed themselves as too sophisticated in their nineteenth- and then twentieth-century scientifically enlightened world to believe that bodies dead for centuries and long since returned to dust could come back to life. So they borrowed, from Christian neighbors mostly, the notion that "the dust returns to the earth as it was; the spirit returns to God who gave it. It is only the house of the spirit which we now lay within the earth; the spirit itself cannot die."[3] I say *mostly* from their Christian neighbors because even early Rabbinic Jews (today we would call them Orthodox), as centuries wore on and the expected messianic end of history was frustratingly slow in arriving, began speaking of the soul's temporary sojourn outside the body while awaiting eventual resurrection when the Messiah arrived. One of the most appealing speculations about where the souls of the righteous await the eschaton has them attending the *yeshivah shel ma'alah*, a heavenly rabbinical academy where they may study Torah with Moses, Aaron, and the great rabbis. To my modern ear this says more about the Jewish love of learning than about literal immortality. No doubt, though, many Orthodox believers still take such speculations literally. The fact remains, however, that, unlike the Christian and Muslim ideas of the soul being freed from its bodily husk, traditional Rabbinic Jews have looked forward to the ultimate reuniting of body and soul, neither of which can realize its destiny without the other. The prayer that I quoted at the outset is a Reform rewrite of a prayer from the classic Orthodox liturgy *Elohai Neshama*, one of the preliminary blessings in the daily morning service, which in its original version begins:

> O my God, the soul which thou gavest me is pure; thou didst create it, thou didst form it, thou didst breathe it into me. Thou preservest it within me, and thou wilt take it from me, but wilt restore unto me hereafter....[4]

Newer non-Orthodox Jewish prayer books tend to omit, or at least leave vague, the notion of resurrection and thus of the separability of body and soul. Another classic Jewish prayer, the second blessing in the petitionary section of the daily service, the *Amidah*, concludes, "Blessed are You, O God, who revives the dead." Reform changed it in Hebrew and English to "who hast implanted within us eternal life"[5] or "the Source of life."[6] The latest Reform prayer book slips resurrection back in as an optional variant reading in parentheses, "who gives life to all" being the dominant reading for which "who revives the dead" may be substituted.[7] I suspect this has more to do with a nostalgia for tradition than with any literal notion of bodies miraculously coming back to life in the future. Conservative prayer books have tended to leave the Hebrew unchanged but similarly to shade the meaning in English with renderings such as "who callest the dead to life everlasting" or "Master of life and death."[8]

An array of ten meditations before the chief mourner's prayer, *Kaddish Yatom*, in the new Reform prayer book, *Mishkan T'filah*—ironically the same book that offers the optional return of resurrection to the liturgy—make no explicit reference to the survival of the individual's consciousness beyond death. Instead, worshipers are guided to naturalistic interpretations of death, such as "With our lives we give life. Something of us can never die"[9] and "those who live no more, echo still within our thoughts and words."[10] Such circumlocutions paper over the confusion in modern Jewish thought. Having questioned immortality of the body, what is to be made of immortality of the soul? More fundamentally, what is a soul that we might each have one? If it is a basic part of each person, why is it absent from the medical books?

HEBREW AND GREEK SOULS

Thanks to the classic Rabbis and the church fathers, the postbiblical notion of the soul as a spirit added to a physical body came to be so taken for granted that interpreters and translators read it into the Hebrew Bible. *Nefesh* and *neshama*, often translated "soul," and sometimes *ru'ach*, "spirit," came to be understood that way. Thus the classic 1611 King James translation of the Bible understood Genesis 2:7 as saying that God "formed man of the dust of the ground, and breathed into his nostrils the breath [*neshama*] of life; and man became a living soul [*nefesh*]." Recent translations are more apt to say "and man became a living being."[11] In context, as the King James translators apparently recognized with *neshama* but not with *nefesh*, God breathed air into the man, not soul, and the man did not become a disembodied soul, but a living person. Similarly, *ru'ach*, often translated as "spirit," also often means "wind." So in Genesis 6:3, when God declares that man is mortal and will not live forever, the King James version translated "My spirit" would not continue in mortals, while the contemporary Jewish Publication Society *Tanakh* says "My breath" will not remain. Modern scholars, even a century ago and overwhelmingly today, have recognized that *neshama*, *nefesh*, and *ru'ach* are all etymologically connected to the movement of air generally, and respiration in particular.[12] As indicators of life, though, as we just saw in Genesis 2:7, sometimes they simply mean "person."

A book could be written teasing out alternative translations of *nefesh*, *neshama*, and *ru'ach* in different contexts. Lacking the ability to hear with biblical ears, as it were, we would still not fully appreciate every nuance. My point is that when we translate these terms as "soul," implying a metaphysical entity, we distort the biblical meaning. Soul was the life that a body had and would eventually lose at death. In the creation story, once God adds ordinary breath to the creature formed from the earth, life could continue through normal procreation, with no biblical claim that God must add either breath or soul to each new body. We can imagine ancient Hebrews sitting with their terminally ill loved ones until they breathed their last breath, at which point they were regarded as dead. So *nefesh*, *neshama*, and *ru'ach* were identified as life force—not synonyms for a metaphysical entity, but indicators of respiration and thus life.

The metaphysical idea of soul came from Greece. Unlike the ancient Hebrews, ancient Greeks drew a sharp distinction between the physical world and the realm of ideas. The physical was perishable and thus, in contrast to pure thought, weaker and inferior. When the Greeks said *psyche* or other terms in the development of their idea of soul, they, too, meant that which enlivens a body. But it was not something so physical as breath, but one or more abstract add-ons "which endow the body with life and consciousness."[13] For Plato, for instance, "the soul, in so far as it beholds the world of ideas, is pure reason. The body is an impediment to knowledge, from which the soul must free itself in order to behold truth in its purity."[14] The soul, in fact, does not learn new concepts from worldly experience, but in a process known as anamnesis is merely reminded by experience of what it knew at its creation and forgot—for how could pure reason be improved by anything so lowly as the material world?[15] It would lead us too far afield to consider all the permutations of mind-body dualism in Greek thought. The point is that this dualism was present in ways that never crossed the minds of Hebrew Bible authors. For Hebrew thought, soul was the enlivening principle of the body that could no more exist without a body than a body could be alive without a soul. We might almost say that for Hebrews, soul was an activity, being alive, a function of a body, not an entity. For Greek thought, the enlivening principle was an entity, and it could exist prior to and thus could continue to exist after its union with a body.

As Hellenism spread and became the dominant culture in the Greco-Roman world, Hellenistic Jews, some of whom became the early Christians, in a cultural milieu that took Greek mind-body dualism for granted as simply the way the world was, read this notion of soul into Hebrew thought. Rabbi Simlai, for example, an early third-century Talmudic sage, taught that the unborn child learns the entire Torah in the womb, but an angel slaps him as he emerges at birth, causing him to forget it all. Whether or not Rabbi Simlai realized it (and he may have), this is Platonic anamnesis dressed up in Hebrew form, even with proof-texts (Talmud, *Niddah* 30b). Although anamnesis did not become basic Jewish thought in later eras, the notion of the soul as a separable entity did. Thus in Rabbinic thinking God did not give Adam life by simply initiating life-giving respiration. God breathed into the perishable matter an add-on soul. And because it was a distinct entity, it might have some existence not only before but also after its sojourn in a body. *Voilà!* Christians had immortality of the soul. Rabbinic Judaism, somewhat Hellenized but less so, resisted the radical separability of the body and soul, but—as we saw in the *Elohai Neshama* prayer—assumed the Bible meant (Greek) life-giving soul and not (Hebrew) life-giving respiration. Eventually, they could even imagine a "life" for the soul in the period between death and resurrection, but they rarely—at least not before classic Reform Judaism in the nineteenth century—thought to utterly separate body and soul. In Jewish tradition prior to Reform, and for non-Reform religious Jews still, personal immortality ultimately had to mean physical as well as spiritual reviving.

As Hellenistic thought was irresistibly influential to Jewish and Christian heirs of the biblical tradition some two thousand years ago, scientific thought seems irresistible, almost self-evidently valid, in our day. Science sees consistent order in the universe, the laws of physics, for instance, but no platonic realm of ideas of which our world is a pale and imperfect reflection. Such Hellenistic ideas strike us as quaint, not as compelling. Ideas may be stored outside our brains—in libraries or on microchips, for instance. But we are learning more and more about brain circuitry. Eric Kandel won a Nobel Prize in Medicine for showing how short-term memory, which is an electrochemical bonding of neurons, becomes long-term memory, which is a physical bonding.[16] The brain, moreover, as we have seen, has specialized circuitry for all sorts of functions, including abstract analysis and planning, which we do so much better than lower animals because we evolved language. This more scientifically sophisticated self-understanding and the presuppositions of the scientific method (such as the need to design and implement replicable experiments before making truth claims) explain why so many moderns, religionists struggling with prayer-book language no less than secularists, are not sure what a soul might be, much less that it is a God-given metaphysical entity that survives the body's death.

The biblical notion of *nefesh/neshama/ru'ach* as life force, even if not a metaphysical entity, fares little better with our scientific worldview. Our medicine does not see breath as the basic indicator of life. Respiration is one of several systems that must operate with reasonable efficiency for us to survive, no more important than blood circulation, nutrition, digestion and excretion, and so on. The Torah, it may be worth noting, also recognized blood as necessary for bodies to live. Blood, too, is *nefesh*—life, not soul—a divine gift (to animals as well as people) and therefore not to be eaten (Leviticus 17:11; Deuteronomy 12:23). The classic Rabbis, moreover, realized that multiple functions were necessary for life and added another prayer for daily morning recitation praising God,

> who formed the human body with skill,
> creating the body's many pathways and openings.
> It is well known before Your throne of glory
> that if one of them be wrongly opened or closed,
> it would be impossible to endure and stand before You.
> Blessed are You, Adonai, who heals all flesh, working
> wondrously.[17]

It would be a stretch to hear any awareness of brain function, the neurological system, in this. The classic Rabbis spoke of good and bad inclinations within us (as we shall discuss in chapter 5) and thus certainly had some sense of thought processes, but that goes unmentioned in prayers acknowledging our dependence upon God for life. The irony for our time is that while many anatomical systems must work for us to live, the one contemporary medicine has come to favor for defining viable life is the neurological system. When the brain can no longer function, even if respiration, circulation, and nutrition can be maintained, we regard a person as legally dead. If, according to the ancient model both Hebrew and Greek, the difference between a body and a living body is the presence of a soul, then it is in the functioning brain that moderns should expect to find that soul.

OUR METAPHORICAL SOULS

We could simply discard the concept of soul as superfluous. Modern medicine, including neurology, can explain well enough how life works without it. Before we do so, however, and discard in the process centuries of significant thought and beautiful poetry about human nature, let us remember that words and concepts evolve over the ages, though echoes of earlier meanings may remain. I suggest that in modern usage the idea of soul long ago became primarily a metaphor for an aspect of consciousness, not the description of a Hebrew enlivening physical process or a Greek enlivening metaphysical entity. We still hear those origins echoed in the wish that a soul should rest in peace or in other religious usages. But when we say that a Beethoven sonata touches the soul, we mean that we not only like it, but we also find it deeply moving. And when African Americans or, by extension, other ethnic groups speak of "soul food," what is meant is food so suggestive of a people's historical roots that it enhances self-awareness and identity. Soul in such cases is metaphorical, pointing to something important in people's consciousness.

Consciousness is such a tremendously important aspect of who we are that we should consider consciousness in general first and then zero in on the aspect of consciousness that most of us today mean when we speak of soul.

The link between the postbiblical Jewish and Christian usage of "soul" and the modern usage, I suspect, has to do with our introspective sense of self. As we think of all the things that our brains consciously keep track of—sensory data, the story we feel ourselves living, planning and decision making, our recollections of the past and hopes and dreams for the future—we develop a sense of "I," "ego," or "self." We are apt to think, intuitively—and Descartes asserted this as a philosophical truth and imagined a miniature self at the central controls—that this must all come together at one spot in the brain that is the self or soul. This is the self that "owns" the mind and the body. When we say "my" body, "my" toy, or "my" anything else, this is the me, the self, the consciousness. When this self is in residence in our brains, we are alive, and when it disappears, we are dead. Because most of us share this sense of there being a center in our heads where everything comes together, we need to account for where we get that idea. Modern cognitive studies argue very convincingly that there is no such single spot, much less a miniature me at the controls, in our brains. First of all, just as a matter of logic this would not solve the mystery of consciousness. Once we were to grant that a miniature me is in there surveying all the data and making the decisions, we would have to account for *its* consciousness and decision-making process. More significantly, as we have discussed earlier, with various sorts of brain scans neuroscientists can show memories being processed in certain areas of the brain, sense perceptions in other areas, language, laughter, aggression, self-control, and so on in still other places. There is simply no one place where it all comes together, and yet (except for a relatively few who are mentally ill) we each have a sense of being a unitary self. It is wondrous—and still somewhat mysterious!

We have already noted that there is a great deal going on in our brains that we are never conscious of and quite a bit going on at any given moment that we filter out in order to concentrate and not be overwhelmed. Daniel Dennett suggests a very useful metaphor to help explain consciousness. When you use your computer to do word processing or to play a video game, there is a user interface. None of us could keep track of, and most of us would not understand, all the processes that are going on at once in the computer. So the computer programmers provide us with a user interface that gives us just the information we need to type, find the virtual files we need, or shoot enemy warriors as we maneuver through the virtual environment of a game. Analogously, for most any task—reading, decision making, enjoying the sights, sounds, and smells of the world around us—far more is going on in our brains than we could consciously keep track of. We have no need, and most of us little ability, to understand the far more complex operations of the brain behind the interface. So our brains evolved to give us that interface, our consciousness. It (this ego/I/me/self) presents far less than what is actually going on in our brain, but it is the part of us we are subjectively aware of. We need it, in a sense, to know ourselves, to know who we are.[18] Consciousness is a virtual self, a user interface that gives us access to and awareness of ourselves as sometimes object of and sometimes cause of what is happening around us. Plants and lower animals have no such consciousness; they don't think about their own thinking. Some higher animals—chimpanzees, for instance—do apparently have some sense of self, but at a far less sophisticated level than humans do.

So, returning to the soul issue, we could simply say that the soul is the religious metaphor for consciousness. That comes close to what Greeks and Hellenized Jews meant by soul, but with a very major difference. Consciousness—to say it again—is not a metaphysical entity that we could today name it soul (in the Greek sense), but one of multiple functions of our brains. Although it is a biological process, plants and most lower animals manage to live without it, so it is not uniquely the life force that we would call it soul (in the biblical sense).

FUNCTIONS ONCE ATTRIBUTED TO SOULS ARE CARRIED OUT BY BRAINS

Conscious thought includes not only rational thought, our manipulation of abstract ideas, but also emotions, the feeling part of our consciousness that we commonly mean by the term "soul." "Soul" is a very useful metaphor, what we are referring to when we talk about sunsets, mountain ranges, or great works of art thrilling our souls, or the bravery of those risking death for a noble cause moving our souls, or, for that matter, the greed of some and the depth of depravity of others depressing or oppressing our souls.

Science is learning how much of this works. A young couple comes in to talk to me about marriage. As they hold hands and stare adoringly at one another, there is no question in my mind that they are physically attracted to each other. And why not?

She has the sort of waist-to-hip ratio that studies have shown men prefer, and he has the broad shoulders and well-developed muscles that studies show women prefer. These figures also indicate generally better child-bearers among women and better providers among men—at least in premodern societies where men hunted and plowed and did so more effectively if they were what we call "in shape."[19] But, I wonder, is there more than physical attraction here? "What attracts you to him?" I ask the woman. "He makes me laugh!" she says. Evolutionary biologists and neurologists have insight to offer there, too. Neurologist V. S. Ramachandran speculates that laughter developed as the body's all-clear signal. There is a potential threat of some sort, something upsetting—a man slips on a banana peel, or a five-year-old asks if you want to hear a dirty joke. If the man breaks his arm, you will not laugh; if he bruises only his ego, you will signal relief to yourself and all around: all clear! You may be embarrassed or appalled if the child says something he cannot (or at least should not!) understand. But if, as you tense up waiting for the joke, he then says, "A boy fell in a mud puddle!" you will realize your nervousness was unnecessary and laugh as much at the situation as at the joke. Says the neuroscientist, "I suggest the rhythmic staccato sound of laughter evolved to inform our kin who share our genes: don't waste your precious resources on this situation; it's a false alarm."[20] Furthermore, from the strictly neurological side, laughter involves "a relatively small cluster of brain structures" in emotion-processing areas of the brain. Brain hemorrhages have been known to provoke hysterical laughter, as has electrical stimulation of an area associated with emotion. You can literally laugh yourself to death, being unable to eat or sleep, from such brain malfunction, though fortunately this is very rare.[21] More commonly, of course, when the brain sends the requisite electrical signals, we laugh—and presumably my would-be bride especially appreciates the happy feeling of well-being her groom keeps reinforcing. The young lady in my office, of course, need think of none of this and would appreciate him neither more nor less if she heard a lecture on the subject.

Why, I query, does the young man find the woman so lovable? "She is so artistic, her piano playing exquisite, and her watercolors so peaceful!" The pleasures such talents produce, some have said, are accidental by-products of our visual and auditory perception. With music, for instance, our brains naturally find patterns and thus develop expectations of the pattern, the tune at the simplest level, continuing. The talented composer sometimes fulfills our expectations and sometimes surprises us, and with other features that can be varied—tempo, loudness, type of instruments (timbre), and so on—conveys mood and emotion, as well. The performers interpret further, playing a bit faster, slower, or louder, holding back a crucial note or crescendo just long enough, a split second, to enhance suspense, or exaggerating a rhythm to get your foot tapping. There are innumerable effects that may delight—or not!—the listener.[22] Visual effects may also involve fulfilling and frustrating audience expectations, and as different sorts of sound play differently on a listener's emotions, different colors and contrasts do on a viewer's. Then again, some insist that such considerations are very much secondary. The arts are a human equivalent to the peacock's tail, which troubled Charles Darwin as an apparent waste of energy for the bird to cart around, while also making the peacock more visible to predators. Flashy though peacock feathers are, they seem dramatically at odds with the idea that only that which provides adaptive advantage survives the evolutionary process. The advantage, it is now believed, for the peacock and the artist, is in advertising one's excellence as a potential mate. Darwin called this "sexual selection." Denis Dutton writes:

> According to this way of formulating sexual selection, an animal shows its genetic fitness to a mate by squandering resources that a less fit animal could not afford to waste:

the endless singing of a mockingbird and the intense red
of a healthy stickleback, not to mention the peacock's tail,
are handicaps, proving, so to speak, that "I can take on
the world with one hand tied behind my back."[23]

My potential groom's infatuation and all the "groupies" flocking
around rock stars demonstrate that this phenomenon operates
in people, too.

How unromantic, we might think. Fear not: each of these
reasons for the couples' mutual attraction is but one of countless aspects of their personalities, situation in life, and values.
They would not likely have ended up in my office had they not
been courting a long time and developing interests, friendships,
and dreams together. Broad generalizations about sexual
attraction, laughter, or aesthetics barely begin to address how
these individuals' unique experience has shaped each of them,
preparing them, in effect, for one another. Yet they should serve
to demonstrate that aspects of conscious experience sometimes
regarded as too mysterious to be accounted for without
recourse to a soul are becoming increasingly understandable
as the normal functioning of our evolved brains. Figuring
out how, neurologically, each of our sensitivities works and
why they might have evolved is a relatively young scientific
endeavor. Theories will no doubt be refined or even radically
changed as we learn more. But it is all happening on normal
neural networks. That does not stop me from telling this couple
that they have been blessed to find one another and fall in
love. How marvelous that the natural order of which we are
a part has prepared us, over thousands of generations, to need
others and find such rich satisfaction in loving relationships.
How wonderful, given each of their complex individuality, that
they have apparently found a good match. The divine order is
manifesting itself in their lives no less than in the beauty of the
heavens. Or, if you prefer the human metaphor for the divine,
God brought them together.

Examples of this sort could be multiplied. The point is that what ancient Hebrews and Greeks thought the soul was doing or what many a modern regards as the activity of the soul can be explained in other ways as brain function. The things we associate with the soul are no less important or profound for that. Most of us would not find life worthwhile without these aspects of our lives. Still, soul is not an add-on to the brain, but rather an aspect of consciousness intrinsic to the brain's normal operation. Consciousness encompasses many routine tasks—the chitchat of ordinary conversation, awareness of our surroundings, our comfort and discomfort, and all sorts of decision making, planning, and organizing—a thousand trivial and not-so-trivial things that we rarely associate with soulfulness. We use the term "soul," then, not for all aspects of consciousness, but for a significant chunk of consciousness. *"Soul" is the collective metaphor for the aspects of our consciousness that we subjectively feel are the essence of our humanity.* "Soul" is a convenient shorthand for our love, our creativity, our aesthetic capacity; for our apprehension of truth and justice; as well as for the opposites of all this—our hatred, cruelty, and greed. "Soul," further, is the metaphor for our yearning, our exaltation, and our suffering and for our faith. Individual differences render an exhaustive list impossible. These tend to be or to evoke emotion and thus, not surprisingly, figure in our spirituality. We certainly need not regard our souls as less important to us for the recognition that they are embodied in our brains as part of consciousness.

WHAT ABOUT IMMORTALITY?

By now it should be clear that the brain is an organ of the body, and our consciousness is one of its functions. So when the brain ceases to function, the soul disappears. I know that is a bitter pill for some people to swallow. Most of us, at least when we are not suffering severely, and some even then, do not want to die. The brain's prime function is to guide us through life, so it (thus each of us) has a will to live. We can scarcely imagine our own nonbeing. When we do, fear of the unknown can produce anxiety. But pain and anxiety are also largely brain-based. As one prayer book puts it, "There is no pain in death. There is only the pain of the living as they recall shared loves, and as they themselves fear to die."[24] There is nothing to fear.

That there is no immortal soul in the sense of one's personal consciousness surviving death—"one short sleep past, we wake eternally," as poet John Donne put it[25]—does not preclude other sorts of immortality but does preclude what is commonly called *personal* immortality. I explained at the beginning of the chapter that biblical Hebrew thought did not imagine that the soul could go on after the body's death; it was primarily under Greek influence that early Rabbinic Jews, and Christians, made that promise to the faithful. They meant well. It was and for many still is a comforting notion. We might explain it in modernist terms as myth, an assertion that God, ruling over all including both life and death, is stronger than death.[26] That still undermines any literal understanding of personal immortality, so we might ask what other form of immortality we might embrace.

The major idea of immortality in the Hebrew Bible is corporate, not personal. Strange though it sounds in our intensely individualistic modern culture, in the ancient world people thought of their identity as primarily a matter of the family, tribe, or nation of which they were a part. The Hebrew covenant with God was a covenant not with individuals, but with the people. The biblical promise of immortality, too, was a promise that the people Israel would endure, not that Abraham or Jacob, Rebecca or Ruth, or any of the prophets would escape death. When Abram first appears in Genesis 12, he is promised that in return for doing what God commands, God will make of him a nation, make his name—his reputation—great, and give land to his offspring (Genesis 12:1–7). In Genesis 15 and 22 he is promised that his offspring shall be as numerous as the stars in heaven (Genesis 15:5) and the sands on the seashore (Genesis 22:17), a promise repeated to Jacob (Genesis 28:14) and quoted subsequently, including by Moses (Exodus 32:13) and Nehemiah (Nehemiah 9:23). As an example not of sweeping promises but of case law, consider the daughters of Zelophehad in Numbers 27. Their father died, they say, of natural causes, with no male heir. "Let not our father's name be lost to his clan just because he had no son!" (Numbers 27:4). To preserve their father's line, they successfully argue, it is only fair that they, though female in a patriarchal society, be given his share of the family inheritance. Being fair to the father in this case required allowing daughters to inherit. They were his immortality.

Admittedly, there are a few biblical passages from which we might infer personal immortality of sorts. Earlier we mentioned the necromancer at Ein Dor summoning Samuel from the grave (1 Samuel 28). Sheol is referred to several dozen times, apparently a shadowy realm of the dead, though perhaps simply a metaphor for the grave.[27] Jon D. Levenson demonstrates convincingly that as the biblical idea of Sheol developed, Sheol came to be regarded as an unpleasant place, "the prolongation of the unfulfilled life," not a place most people expected to inhabit after death. Those who died fulfilled looked forward not to personal immortality, but to continuity through progeny.[28] Even if one were to take the relatively rare hints of personal immortality in the Hebrew Bible as evidence of a widespread belief, the alleged biblical view would be as untenable in our time as postbiblical Rabbinic, Christian, or Muslim promises. Personal immortality is wishful thinking.

Corporate immortality, on the other hand, an important biblical idea, makes as much sense today as it did in the past. We live on through our children. At least as important, particularly because some do not have children, we live on as part of our people. This still resonates for Jews after several thousand years of history and could easily be adapted by other groups. The Bible does not claim God cares only for the people Israel, but for other nations as well. Israel is no better than the Ethiopians, and God has liberated Philistines and Arameans, not only Israel, said Amos (9:7), and Isaiah 19:25 has God declaring, "Blessed be My people Egypt, My handiwork Assyria, and My very own Israel." Each of us can be part of a religious or national tradition or, for that matter, a tradition of science, art, or any other human endeavor, contributing to something that goes on past our own death. Many of us take great satisfaction in that and find that it lends added meaning to our life struggles. Though few of us will personally be remembered for more than a few generations, our efforts add to the "chain of tradition." To put this differently, there is an element of self-transcendence when we devote ourselves to groups and activities that are larger than ourselves. This is a less self-centered notion of immortality. Our souls themselves do not go on, but much that most profoundly concerned and motivated them does.

5

Free Will and Free Won't
Programming Your Brain

Individual responsibility and dignity are predicated on free will. Free will is threatened by the scientific understanding of conditioning and of will, the latter an "ownership emotion" that alerts the brain to its responsibility for a decision after it is made. Still, because of the human gift for language and the abstract thought language enables, we—and only we among living creatures—have a good deal of ability to program our own brains. Thus we have significant, though not absolute, freedom to make decisions. Classic Christian and Jewish ideas of, respectively, original sin and *yetzer* psychology need refinement in response to all this.

> *A smart aleck decided to show everyone that the sage they held in such high esteem was not so smart after all. He approached with a sparrow in his hand behind his back and asked the sage, "Is this bird alive*

*or dead?" If the sage responded, "Alive," he would
crush it. If the sage said, "Dead," he would open his
hand and it would fly away. The sage responded, "The
answer is in your hands."*[1]

You are at a dinner party. You have eaten a big meal. You are not hungry. Then the hostess appears with a gorgeous cheesecake. (If you don't like cheesecake, substitute *your* favorite.) Perhaps, like me, you grew up with the rule that you could only have dessert if you had cleaned your plate, which delivered the more subtle message that if you had finished the meal you had earned dessert. But now you are an adult and worried about your waistline or perhaps your cholesterol or diabetes. Parties are for enjoying, and you would not want to offend the hostess, you rationalize, so you enjoy a generous slice of cake, which is as wonderful as it looked. Could you, exercising free will, have turned it down? Because you obviously like it so much, your hostess says, you should have another piece! Do you have the ability, the free will, to turn it down?

The societal consequences and thus the seriousness of such decision making escalate dramatically if we consider more emphatically ethical examples. A certain remarkably attractive person who is married, but not to you, gives you a look, or makes a remark, or both, which seem to be saying, "I'm sexually available. Interested?" Maybe you are married, too. Either way, "Are you interested?" is scarcely the right question. Tens of thousands of years of evolution have genetically programmed your interest in sex. But chances are that as a matter of personal morality you believe that promises should be kept, love honored, and spouses, even another's and certainly your own, should not be betrayed and hurt. As you were growing up, you learned that society frowns on marital infidelity, and if you have any religious education, you may well believe, based on the Ten Commandments and other sources, that adultery is a sin against God (Exodus 20:13; Deuteronomy 5:17). The germane question, if you agree the sexual encounter would be wrong, is whether you have the ability to overcome temptation.

The pragmatic consequences of resisting or giving in to temptation in such cases are obvious. The cheesecake may damage your health. Adultery may undermine your marriage and family, the marriage and family of the other, and ultimately the stability of family-based society. Especially with adultery, your reputation may be damaged if the affair becomes known, and self-respect may be compromised, as well. Short-term pleasure should be weighed against larger and longer-term consequences. Theologically, psychologically, and even legally the issue is even more profound. Human dignity is premised on human freedom. We have all sorts of physically and psychologically based needs and desires. Those most directly relevant to the above examples include the need for food, sex, pleasure, ego gratification, self-respect, and the respect of others. If our physical or psychological cravings are so strong that we lack the power to resist them, we are robots, very complex robots to be sure, but still automatons acting out our programming, not independent agents making decisions. In that case there is neither shame in succumbing to temptation nor nobility in resisting, no credit due for right behavior or blame for wrong. We do not think ill of a toddler for grabbing another child's toy. Self-control, respect for the rights of others, and then sharing must be learned. Even among adults, our society will occasionally find someone accused of a crime "not guilty by reason of insanity," which we often define as "not knowing the difference between right and wrong." There is no moral culpability—or its opposite, rectitude or nobility—unless we first know standards and then have the ability to act on them. Whether or not we have free will is a high-stakes issue.

The Hebrew Bible largely assumes human free will. The archetypal man and woman are easily led astray, but not for lack of understanding God's instruction or having the ability to obey. Eve argues with the snake, showing she understands God's instructions. After the sin, Adam and Eve hide from God, and each tries to pass the blame off on another, revealing that they know they have done wrong. Adam and Eve are not reliably good but are nonetheless capable of goodness, else the story's audience, ancient and modern, would not feel that punishment is justified. Their sin shows they are deeply flawed. Their punishments, nonetheless, reveal that God regards them as free and thus culpable. One of the points of the story is that the temptation to sin is real, but should be resisted (Genesis 2:4–3:24). That message is even more explicit in the next Genesis tale. God sees Cain is upset that Abel's offering is regarded as superior to his own. So God analyzes Cain's situation for him:

> Cain was much distressed and his face fell. And the
> Lord said to Cain:
> "Why are you so distressed,
> And why is your face fallen?
> Surely, if you do right,
> There is uplift.
> But if you do not do right
> Sin couches at the door;
> Its urge is toward you,
> Yet you can be its master."
>
> (GENESIS 4:5–7)

Cain's upset, even perhaps his urge to kill, is understandable. But he does not have to act on his base inclination, and neither, the Torah implicitly argues, do we. Sin tempts, "yet you can be its master."

As Genesis continues to explore human nature, the message of the Noah story is still darker. God resolves to destroy the world because "all flesh has corrupted its way," animals and humans alike. After the flood, God pledges not to destroy the world again, realizing that no better can be expected of us, for "the devisings [*yetzer*] of man's mind are evil from his youth" (Genesis 8:21). In other words, total destruction would be unfair because we are so flawed. Some sin, to be sure, we can resist, but never consistently enough to be wholly meritorious. The "devisings of man's mind" are powerful enough that though God can expect some goodness from us, some sin is inevitable. We are neither hopelessly bad nor reliably good.

Divine commandments in general, and in particular such exhortations as this from Deuteronomy, "I have put before you life and death, blessing and curse. Choose life—if you and your offspring would live" (Deuteronomy 30:19), would be unnecessary if the biblical authors thought we would always choose rightly, and nonsensical if they did not think we had the ability to choose. Likewise, numerous prescribed atonement sacrifices presuppose that we will sin, and importunings to "cease to do evil; learn to do good" (Isaiah 1:16–17) assume our capacity to do better. In the Jonah story, for instance, we see that even the worst of sinners, those cruel and rapacious empire builders the Assyrians, have the capacity to repent, turning away from evil. Even in the Exodus story, where God says in advance that God will harden the heart of Pharaoh (Exodus 4:21), Egypt and its symbolic god/ruler deserve punishment for oppressing the Hebrews prior to God finally acting. Before God intervenes to harden Pharaoh's heart, "Pharaoh's heart stiffened" on its own, and he does not let the people go. While God does harden Pharaoh's heart several times, including before the tenth and worst plague, Pharaoh arguably deserves the punishment because he changes his own mind ("stiffened his heart") repeatedly (Exodus 7:13, 7:22, 8:11, 8:15, 8:28, and 9:7). Not until Exodus 9:12 does God stiffen Pharaoh's heart, and Pharaoh is back to making his own wicked decision by 9:35. In sum, the biblical authors believed that God's commandments and warnings could help us to be good, and we can thus be held accountable for our actions. Yet our nature is such that we can never be wholly good.

A tension, if not quite a contradiction, remains. We are both capable of choosing rightly and vulnerable to choosing wrongly. This is an ironic predicament. We have some free will, just not enough to be as good as we think we ought to be. This begs the question: if we cannot be perfect, could we not have been created—or evolved—at least a little *better*?

It is significant that individuals can learn from making mistakes. "See how happy is the man whom God reproves; do not reject the discipline of the Almighty" (Job 5:17). The people as a whole, likewise, should learn from punishment, as in Jeremiah 2:19:

> Let your misfortunes reprove you,
> Let your afflictions rebuke you;
> Mark well how bad and bitter it is
> That you forsake the Lord your God,
> That awe for Me is not in you
> —declares the Lord God of hosts.

So, once again, we find freedom and educability assumed. But beyond asserting that God created us, the biblical authors, never philosophically inclined, apparently felt little need to explain how our wills work as they do.

A CLASSIC CHRISTIAN ATTEMPT TO SOLVE THE PROBLEM OF FREE WILL

Hellenistic Judaism brought Greek mind-body dualism to the problem of explaining why, though created by God, we are flawed and sinful. We have the urge to sin because of our lowly physical nature as well as the capacity to rise above that base

aspect of the self and be good because of the rational faculty—
soul and thought, purely spiritual rather than physical—that
is equally a part of our nature. Hellenistic Jewish philosopher
Philo Judaeus in first-century Alexandria, for example, argued
that animals cannot sin, for they are devoid of reason, and
heavenly creatures do not sin, "not being bound up in the
region of interminable calamities, that is to say, in the body."
But "man," knowing good and evil but burdened with the
corrupting influence of physical bodies, "often chooses the
worst" and is "most justly condemned as being guilty of
deliberate and studied crime."[2] Take the biblical idea that we
are—for some unspecified reason—morally flawed, read it
through the lens of Greek mind-body dualism, and it seems
obvious that the flaw must relate to our physicality. Thus in
Christianity, the most influential Hellenistic Jewish synthesis,
one finds the doctrine of "original sin." The apostle Paul, in
the New Testament, wrote, "I see another law at work in the
members of my body, waging war against the law of my mind
and making me a prisoner of the law of sin at work within
my members" (Romans 7:23). There would be much debate
among early church thinkers about what Augustine and others
called "original sin." In Genesis, Adam and Eve, as the first
people, were the first sinners. But original sin as a doctrine
transcends Adam and Eve's disobedience in the Garden of
Eden, referring back to its cause, human failure to overcome the
negative potential of what Augustine called "concupiscence,"
sexual desire, and forward to its consequence, physical death.
Although Augustine's mind-body dualism was not total, for
properly channeled sexuality was not inherently sinful, original
sin was an aspect of human nature, not a single act. Passed on
through the generations through sexual intercourse, it involved
a negative view of sexuality and thus physicality.

In discussing soul in the previous chapter, we saw that
mind-body dualism was not part of biblical Hebrew thought.

Yet if one starts—as many a Jew or Christian in the Greco-Roman world did—with the idea that mind-body dualism is simply the way things are, a self-evident description of reality, then explaining sin this way is logical. Original sin is, from this rabbi's perspective, a reasonable Christian midrash, a way of reconciling and bringing to life for residents of the Mediterranean world contemporary Hellenistic thought and the Hebrew biblical stories and worldview. For Adam and Eve, and then equally Cain, Noah's generation, and others down to and including us, sin involves the appetites of the flesh. The path to righteousness is learning to control our appetites, and the problem of our dual nature will finally be solved only at death, when the soul is freed from the prison of the body.

From this Hellenistic perspective, then, the unaided will is not strong enough to resist bodily hungers. Even with exemplary efforts at self-discipline, will is not adequate to the task. So God provides guidance in the form of scripture and community, and more pointedly in Christianity, God sent God's son to atone for people's sins. To accept that gift and overcome one's sinful physical nature, the soul—the spiritual as opposed to the physical aspect of the self—must be fully entrusted to God. It will then be pure enough not to be dragged down to the grave with the body it temporarily inhabits. "For God so loved the world that He gave His only son, that whoever believes in Him should not perish but have eternal life" (John 3:16). The body will still be subject to death, but the soul, having transcended the body's sinful nature through faith, will be worthy of eternal life. Why did God provide this wondrous escape from the intrinsic inferiority of physical existence? God just did. God's love is an act of grace, unmerited blessing.

This is a significant part of classic Christianity, one of those higher paradigms of which I spoke at the end of the first chapter. For those who see the world through this lens, it was and remains a powerful way to make sense of a large chunk of experience. Original sin, though, tied as it is to sexuality and physicality, has needed reinterpreting (or ignoring) in more modern times, because mind-body dualism no longer seems so self-evidently valid.[3] Teaching Jewish-Christian dialogue in a liberal Protestant seminary and suggesting that Jews and Christians differ over original sin, I have found time and again, initially to my surprise, that this classic Christian doctrine is incidental at best to the theologies of many students. Because it remains official doctrine in most denominations, though, I must observe that, to this non-Christian, the classic Augustinian doctrine of original sin appears to compromise free will. Although God's grace enables individuals to exercise their free will to overcome their natural sinfulness, on their own they could never do so. Only believers, therefore, not all people by their very nature, can fully exercise free will.

A CLASSIC JEWISH ATTEMPT TO SOLVE THE PROBLEM OF FREE WILL

As early as the time of the Mishnah, compiled in the second century CE, Rabbinic tradition began to develop a very different approach to explaining sin. As opposed to the Hellenistic philosophical approach, Rabbinic Jews developed a psychological model of the self. First- to second-century authority Rabbi Joshua observes that the *yetzer hara*, "evil inclination," is one of the things that leads to death (or perhaps to the inability to function in society; evil inclinations "drive a man out of the world" [*Mishnah Avot* 2:11]). Second-century sage Ben Zoma identifies the strong man or hero as the one who "controls his *yetzer*" (*Mishnah Avot* 4:1). This echoes the use of the term *yetzer* we saw above when God, after the flood, promised not to destroy the world again because "the devisings [*yetzer*] of man's mind are evil from his youth" (Genesis 8:21). As we shall see, later rabbis in the Gemara (codified in the sixth century as a commentary on the Mishnah; printed together they comprise the Talmud) and midrashic collections continue to expand and expound this understanding of the origins of human sin. Various hungers that often get us into trouble result from the influence within each person of this "evil inclination." We see its power when we allow our desires, especially our lust for sex, wealth, power, fame, and pleasure, to dominate us. In creating us, though, God balanced the *yetzer hara* with a *yetzer tov*, a "good inclination." God further graciously gave us Torah, which, properly interpreted, can even give us answers to new issues that arise. If you think back to the cheesecake temptation with which I began this chapter, you have surely had an internal dialogue along the lines of "One more piece won't hurt!" "Yes, it will; you promised to watch your weight!" "You can diet or do more time on the treadmill tomorrow!" "Maybe, but what about my cholesterol?" So, in situation after situation—the adultery example could provide an even juicier discussion—the *yetzer hara* and the *yetzer tov* do battle within each of us time and again.

The first objection to this is raised by the Rabbis themselves. Does not saying God created the *yetzer hara* entail imputing evil to God? Actually, no. The Sages, in the midrash on the sixth day of creation in Genesis (1:24–31), insist, "But for the *yetzer hara*, no man would build a house, take a wife, or beget children" (*Genesis Rabbah* 9:7). Referring to Ecclesiastes 4:4, "I have noted that all labor and skillful enterprise come from men's envy of each other," they add that building businesses is also a positive result of the so-called *yetzer hara*. *Illicit* sexuality or business dealings may get us into trouble, but God planted desires in us so we would build, marry and procreate, and work at our businesses. Do these activities properly and they lead to blessing, not sin. By extension, all the desires we associate with the *yetzer hara* are good. There is nothing wrong with accepting God's blessings, including both fleshly pleasures and psychological satisfactions. When we go to excess, the fault is ours, not God's. The human need is to remain ever vigilant to keep the influences of the *yetzer hara* and the *yetzer tov* in righteous balance.

In our context, another objection must be raised. The Rabbis assume some third entity, "you," as part of this psychological model. *You* listen to the opposing inclinations and decide which to act upon. Cain, we just saw, was told sin was tempting him, but "you can be its master." Who is the "you" who makes that decision, and how does that work? Like Descartes' imagined *homunculus*, a miniature self at the brain's controls getting all the input and deciding what to do, this approach could lead to an infinite regress. You "hear" the *yetzer hara* urging you to sin and the *yetzer tov* urging you to resist; perhaps you even study some Torah and do good deeds, activities commended by the Rabbis as therapy against temptation ("Blessed are Israel, as long as they are devoted to the study of Torah, and works of loving-kindness, the evil *yetzer* is delivered into their hands" [Talmud, *Avodah Zarah* 5b]), and still, occasionally at least, this "you" fails to resist temptation. Why, if we genuinely have the power to resist, does no one ever wholly succeed? In Day of Atonement liturgy, Jews recite a confession of sins that includes, "We are not so arrogant and stiff-necked as to declare before You, our God and God of all ages, we are perfect and have not sinned; rather do we confess: we have gone astray, we have sinned, we have transgressed."[4] The Rabbinic theology that urges us to be strong, defining the strong person as the one who controls his evil *yetzer* (*Mishnah Avot* 4:1), admits that no one is strong *enough* to do so all of the time.

Is this not a flaw in God's design? Several rabbis suggest that God regretted the design.[5] But I doubt most would have regarded this as a flaw. God designed us to be constantly at war within ourselves and commanded us to choose rightly: "Choose life—if you and your offspring would live" (Deuteronomy 30:19). That is, we are choosers by nature; our ability to sin or do right is the result and functional definition of our free will. What *seems* to be our design flaw is the mechanism of our freedom. Always choosing correctly might not be true freedom but simply the acting out of a perfect nature. We must be imperfect to exercise free will. But could we not, even as we occasionally sin, do *better*, perhaps with a weaker *yetzer hara* or a stronger *yetzer tov*? There *is* a way to do better. God, as an act of grace, gave us Torah. Torah study strengthens the *yetzer tov*, the Rabbis believed. "As long as you occupy yourselves with it, the *yetzer hara* will not have dominion over you," say the Rabbis, using God's warning to Cain as prooftext: if we do well, which they read as the meritorious act of study, "there is uplift," but if not, "sin couches at the door" (Talmud, *Kiddushin* 30b). It is not, then, as in the Christian model we just examined, that God saves us from sin, but rather that God gives Torah and its commandments so that we can save ourselves.

The stress here, to be sure, is on what is going on in the individual's mind. But a Talmudic parable demonstrates that though the generations that produced this psychology were aware of the mind-body dualism of the Greco-Roman world they inhabited, they regarded the *yetzer hara* and our proclivity to sin not as based solely in body or in soul, but in both. In one version the story is presented as a rabbi's refutation of a Roman emperor insisting that body and soul can escape judgment:

> Antoninus said to Rabbi: "The body and soul can both free themselves from judgment. Thus the body can plead: The soul has sinned, [the proof being] that from the day it left me I lie like a dumb stone in the grave [powerless to aught]. Whilst the soul can say: The body

has sinned, [the proof being] that from the day that I departed from it I fly about in the air like a bird [and commit no sin]." He replied, "To what may this be compared? To a human king who owned a beautiful orchard which contained splendid figs. Now he appointed two watchmen therein, one lame and the other blind. [One day] the lame man said to the blind, 'I see beautiful figs in the orchard. Come and take me upon thy shoulder, that we may procure and eat them.' So the lame bestrode the blind man, procured and ate them. Some time after, the owner of the orchard came and inquired of them, 'Where are those beautiful figs?' The lame man responded, 'Have I then feet to walk with?' The blind man replied, 'Have I then eyes to see with?' What did he do? He placed the lame upon the blind man and judged them together. So will the Holy One, blessed be He, bring the soul, [re]place it in the body, and judge them together...."

(TALMUD, *SANHEDRIN* 91A–B)[6]

The psychological approach of the Rabbis preserves a significant measure of free will and thus responsibility, without denying the inherent flaw in human nature. Thus the good person is not defined as perfect. The Rabbinic standard for goodness is expressed through the metaphor of a balance scale. The good deeds on one side should outweigh the bad deeds on the other. That is the best flawed humans may hope for.

I would note in passing that medieval religious thinkers gave much thought to the logical puzzle of how we can have free will if God is omniscient, knowing all things timelessly, future as well as past and present. "Everything is foreseen, yet free will is given," Rabbi Akiva said (*Mishnah Avot* 3:19), but the medievals recognized a logical problem. If, in the epigraph to this chapter, for instance, God knows what the sage will answer and whether the smart aleck will crush the bird, they may each have the illusion of deciding for themselves, but because God cannot be wrong, they literally cannot do other than what God knows they will do. His response is the only answer the sage can give, and the smart aleck literally cannot do the opposite of what God expects without making God wrong, which for this philosophical approach God cannot be. Much has been written on this. For our purposes it should suffice to say that other personal-God concepts do not all assume such thoroughgoing omniscience. In the Hebrew Bible, for instance, God goes to check whether the residents of Sodom and Gomorrah are as bad as God has heard (so God does not already know) (Genesis 18:21). God tests Abraham and Job (Genesis 22 and Job 1–2). Earlier we mentioned prophetic warnings and promises: change your sinful ways to change your future. All this assumes that God does not know the future, leaving it open. Perhaps God could know or otherwise force the future but as an act of voluntary self-limitation chooses not to do so. Then again, God can have huge power without having the power to make the contingent (what might happen) noncontingent (what must happen). Philosophically this is akin to other logic puzzles, such as "Can God make a weight so heavy that God cannot lift it?" It only leads to incoherence to insist that God can do the self-contradictory.

Classic Jewish and Christian approaches to human nature and free will attempt to speak more directly to our life situation. They see us as flawed, albeit for different reasons, and both offer paths to overcome the flaw—giving full faith to Christ, or studying Torah and disciplining oneself to have the good inclination overcome the evil inclination. Each, moreover, advises participation in a religious community that reinforces its path socially and provides intellectually what I keep calling a higher-order paradigm as guidance for living. All this addresses but does not resolve the paradox: if we make some right choices, but our nature is such that we not only do not but also cannot always choose rightly, what sort of freedom, if any, do we have?

THE TIMING OF OUR CHOICES

Most of our decisions *seem* consciously and freely made. Are they? The Rabbis were on the right track with their recognition of inner dialogue, though they had no answer to who or what the "you" is that decides between the different inclinations. The question never occurred to them. They may have thought of the *yetzer hara* as a single voice within, in which case neuroscience must declare them simplistic. It is at least arguable, on the other hand, that they recognized *yetzer hara* as a collective voice composed of various basic desires (without it, as we saw, "no man would build a house [or business], take a wife, or beget children") (*Genesis Rabbah* 9:7). Classic Christianity's equivalent might be seen as the seven deadly sins: wrath, greed, sloth, pride, lust, envy, and gluttony. Freud called it all "id." This is the looking-out-for-number-one part of us without which we would not flourish, but which, undisciplined, gets us into trouble. Similarly, the Rabbis likely thought of the *yetzer tov* as a single internal voice we would today simply call "conscience." Yet just as we tease out different strands of selfish inclination, it seems likely that there would be multiple mental mechanisms at work in the good inclination, some dedicated to keeping the selfish side from going to excess, and others, more positively, to motivating good deeds. In any event, I began with biblical and classic postbiblical approaches in order to stress that the science to which we now turn is providing further insight into questions we have been struggling with for millennia. We now turn to cognitive studies for a look at what the will and the conscious self, which seem to us to be deciding, actually are and how much room for free will a scientific understanding of decision making allows.

We have seen that large numbers of processes are going on simultaneously in our brains. The brain, and especially its higher-reasoning facility, with input from all the rest—perceptions, hungers, memories, hardwired or learned values, hopes and fears, emotion-enriched thought and thought-guided passions—generates and considers options[7] and then directs what it concludes are our best moves, not only moment to moment but also as longer-term plans. As mentioned in chapter 3, people have far more higher-reasoning ability than do lower animals. The brain regions where this happens, having developed full capacity late in the evolutionary process, are richly connected to earlier, "lower" areas, both to receive input (lion approaching; danger!) and to send action directives (implement coordinated muscle action: run!). Animals do that, too, but people are able to "translate" our thoughts and feelings into language and then manipulate them in that form (a philosophy of handling lion threats: if the lion has seen you, do this; if not, do that; if in this or that terrain ...; if alone or if with others ...). This wondrously complicated process usually seems effortless to us. We decide what to do, which is to say we walk here or run there, write a letter or pull a trigger, resolve to get more information, think further ... or whatever. The issue is not do we decide, but are the choices free?

The brain, no one doubts, is the central processor, albeit with rich input from elsewhere in the body. The brain's frontal cortices do much of the planning, not because everything is transferred there, but rather because things going on elsewhere are linked there so that coordinating and further processing can be done. In other words, there is no "you" spot, no virtual little man at the controls. As multiple processes go on, the brain generates something quite amazing: your consciousness, the sense that you are watching or listening in on your own thoughts and your own decisions and their consequences, the subjective sense that there is a thinking, feeling, deciding, and acting "you."

Once again, insects, rodents, and countless other creatures take in information, process it, and make choices, too. But few if any—probably only our closest simian relatives, and they at a primitive level for lack of our linguistic sophistication—have any sense of thinking about their own thinking, of consciously rather than automatically making choices. They react to their environment based on hardwired instincts and learned strategies. All or nearly all (again excepting a few apes) decisions of lower animals are entirely conditioned. Only we at least seem to have the ability to act against our instincts and the preponderance of our conditioning. If we can do that, even though we do not always do so, then we have free will. How, then, do decisions happen? How does will fit into the consciousness that seems to us to be making them?

For several decades into the early 1990s, Dr. Benjamin Libet, a neurophysiologist at the University of California at San Francisco, experimented with the timing of decision making and consciousness. Sensors were attached to people's heads to indicate when brain activity commenced. Thus wired, subjects were asked to do tasks such as wiggling a finger or flicking a wrist whenever they wanted. They were asked to note the position of a dot moving fast enough on a clock face that the experimenter could measure how many milliseconds elapsed between the reported time of decision, the brain wave activity, and the action. It became clear, not surprisingly, that there was brain activity to initiate the action. But the time subjects reported that they had made the decision was actually 300 to 500 milliseconds *after* the brain activity. In other words, the conscious mind could not itself have made the decision, because the activity was initiated as much as half a second before the subject became conscious of it. Subjects felt as if they had willed the action, but in effect, the conscious mind was late to the party. Where is free will if, before you *consciously* decide, the brain has initiated the action?[8] Libet could thus distinguish in the laboratory between the decision to move, the awareness of the decision, and the actual muscle movement. What seems instantaneous and simultaneous to us involves as much as a full second, the time it takes for electrical signals to be sent and muscles to move. The consciousness occurs only about 150 milliseconds before the action. In that very tiny fraction of a second, you can still counteract the decision, "change your mind," we would call it, leading Dr. V. S. Ramachandran to quip that you may not have free will, but you at least have "free won't."[9]

This may be easier to grasp if you think of professional-level tennis. Daniel Dennett reports that Venus Williams's serves average 125 miles an hour, which means the opponent has 450 milliseconds, less than half a second, not only to hit the ball back but also to decide how and where to hit it back. The normal subject in a laboratory takes 350 milliseconds to see a light come on and then press a button. Clearly, if Williams's opponent has to stop to think about her decisions, the ball will be by her before she can act. The 100 milliseconds between response time and available time is enough to hit the ball back, though, if, prior to the serve, the opponent has thought about options and so, in effect, is implementing a preconceived plan. A new plan is not made with each serve, but rather the well-prepared athlete has practiced her moves again and again for years. A new visual stimulus is reacted to with each serve, but reacted to in accordance with a conditioned pattern. "The tennis player pre-commits to a simple plan and then lets 'reflexes' execute her intentional act."[10] Could she change her mind during the process? If, in baseball, you have ever started to swing at a pitch and then wanted to check your swing, you will be aware of how difficult that is. But it can often be done, and no doubt done more consistently by well-practiced athletes. Is all this a matter of free will or of conditioned reflex? Some of both!

WHAT IS A WILL THAT IT MIGHT BE FREE?

All this is helpful but continues to beg the question of what that conscious you, your conscious*ness*, is. Is your conscious self making decisions or learning about them after the fact? At the very least, Libet's work demonstrates that the latter is often the case. Harvard psychology professor Daniel M. Wegner provides a very helpful model of what conscious will is and a theory of why it evolved that way.[11] After considering Wegner, we will be in a position to make some generalizations about free will and the theological questions with which we began the chapter.

Based on our learning and conditioning over the years (and probably some hardwired predilections, too, as we shall see in the next chapter), we are able to do lots of things—simple things like scratching one's nose, more complex things like driving a car or writing a novel, morally fraught things like telling the truth or lying, helping or killing. All of them require multiple brain processes. We could not consciously keep track of it all; there is too much, and it goes on at such speed that it seems instantaneous. We have already spoken, in the previous chapter, of Dennett's image of consciousness as a user interface, giving us just enough of what is going on inside our heads that we can appreciate it without being overwhelmed. So as the brain makes (some of its) decisions, it gives us—projects, as it were, as if on a screen, but a mental screen—a feeling of authorship. My arm did not just rise to scratch my nose or move rapidly out to punch yours; my brain issued the orders to make that happen. All the factors that went into deciding and implementing do not have to be projected onto this screen of consciousness, just the awareness that the event happened and the sense of authorship. I wanted to scratch my nose, and did, or to punch you in the nose, and did. This is analogous to my computer monitor, which does not do the computer processing but might be mistaken for doing so, because that is what I actually view while the real work goes on elsewhere.

Apply this to Libet's wrist flicking. The experimental subject thought he was first consciously deciding (step 1), then mentally issuing the order (step 2), and then wagging his finger (step 3). In fact he was deciding (step 1) and issuing an order (step 2), only then becoming conscious that the decision had been made (step 3), and finally acted (step 4) and saw the process confirmed by the action (step 5). The conscious you, which has the sense of willing the action, is a secondary phenomenon—the monitor—easily mistaken, in fact intuitively mistaken, for the primary process. Lest there be misunderstanding (for this is, indeed, counterintuitive), we should note one more time that there is no "you" in your head *watching* that monitor. The metaphor breaks down here: the conscious you is the monitor itself, the monitor "watching" itself, aware of itself, as "deeper" processes in the brain project the image of thought going on. The sense of being causal agents is part of consciousness. But the consciousness is not the cause, it is the projection of a simplified version of the thoughts that precede it. Neither is the will the cause, but rather a *feeling*—Wegner calls it "a kind of authorship emotion"—added in so we will know, and feel and remember more strongly than we might otherwise, that we initiated an action or idea:

> In the same sense that laughter reminds us that our bodies are having fun, or that trembling alerts us that our bodies are afraid, the experience of will reminds us that we're doing something. Will, then, makes the action our own far more intensely than could a thought alone.[12]

This at first appears to be bad news for those of us who want consciousness to freely will, to direct, decisions. The consciousness is secondary to the real decision making of the brain. Still, what you "see" on the monitor is genuinely a projection of activity a split second before in your brain. You can reasonably think of this as one process with two parts, the complex thinking and the simplified awareness of what is being thought. The split second between is there because thoughts are not really instantaneous, but happen in real time, and the projection takes a moment. But the projection is genuinely a small portion of your brain process.

That is a little better. You—your consciousness—are not late to the party; you *are* the party! Consciousness is only part of the process that is your mental self, but it is all you. However, this process is still short of your will causing the action, because unlike other aspects of consciousness, the will part is not projected from what was going on initially, but added as a feeling afterward. Wegner explains that when you have a thought that becomes conscious and immediately afterward what you thought about happens, you intuitively think there was cause and effect: you caused the action. Most of the time you did, but not always. Wegner reports on a family visit to a toy store:

> While my kids were taking a complete inventory of the stock, I eased up to a video game display and started fiddling with the joystick. A little monkey on the screen was eagerly hopping over barrels as they rolled toward him, and I got quite involved in moving him along and making him hop, until the phrase "Start Game" popped into view. I was under the distinct impression that I had started some time ago, but in fact I had been "playing" during a pre-game demo. Duped perhaps by the wobbly joystick and my unfamiliarity with the game, I had been fiddling for nothing, the victim of an illusion of control.[13]

The game player thinks of himself as causing the movement. But actually the brain does not feel the causation but adds it as analysis when movement quickly follows thought. Similarly, Wegner invites the reader to imagine looking at a distant branch and thinking it should move. Then it does move. You try this again, and yet again, and each time the branch, which had been still, moves. You are very likely to get the idea that you have somehow, by force of will, caused the branch to move. "Move, branch," you think, and it moves! Then you stop trying to move it, and a minute later it moves again. Probably the wind moved it. Certainly something did (maybe an unseen squirrel?), but not you. Your brain, because the thought and movement came so close together the first several times, mistakenly got the idea that you willed the motion.[14] The cause-and-effect assumption of one thing immediately preceding another gave you the very genuine feeling, false though it was, that you willed something you did not will. Most of the time your brain truly is the author of what it wants to happen next that then happens. But not always. Will, thus, is not something in your brain exercised to make something else happen, but a feeling of authorship your brain adds to the experience after the fact as you construct your world and construe its meaning (which we talked about back in chapter 1).

Why would such a mental ability, adding a feeling of responsibility to consciousness, have evolved? Wegner answers:

> One of the key reasons for describing actions and ascribing them to persons is as a way of determining who deserves what. Action descriptions mark up the flow of human events into convenient packages ("He repaired my motorbike," "She shot the winning basket at the final buzzer," "They sang a musical program at the retirement home"). But each such description of action comes with an ascription, a notation of who is author. Although it could be useful to all of us to remember all the various ascriptions, it is most pressing that we remember our own. We must remember what we have done if we are going to want to claim that our actions have earned us anything (or have prevented us from deserving something nasty). Although the whole team might secretly want to lay claim to that final basket that won the game, for instance, the player who made

the shot would be particularly remiss if she didn't know, or later forgot, that she did it. It is good that when she did it, she had an emotion, an experience of conscious will, that certified it as her own.

Conscious will is particularly useful, then, as a guide to ourselves. It tells us what events around us seem to be attributable to our authorship. This allows us to develop a sense of who we are and are not. It also allows us to set aside our achievements from the things that we cannot do. And perhaps most important for the sake of the operation of society, the sense of conscious will also allows us to maintain the sense of responsibility for our actions that serves as the basis for morality.[15]

OFTEN YOU CAN SAY NO

Go back, for a moment, to the cheesecake and the adultery, and the interior dialogue as you decide what to do. Your brain was doing a lot more than having that dialogue, but a split second before each step of it you genuinely had each thought. It is as if there were a deeper you doing the thinking and a shallower consciousness that had the sense that it was doing the thinking. That the decision was conscious was an illusion, because it happened a split second before your consciousness of it. And the *feeling* of will was added after the decision, too. Still, your brain made the decision. It did so based on your previous knowledge and conditioning.

But was it a *free* choice? How could it be if it happened before you were aware of it? If conscious awareness were instantaneous rather than a split second later, I suggest, nothing would change. You would be reacting to the same external stimuli, considering the situation in the context of your memories, filtering possible action and meaning through the same paradigms, ultimately accepting one of your options. The decision, in other words, is in every respect your brain's, which means yours. But was it so conditioned by previous experience and values that you *had* to decide that way? If so, it was not free. If not, we might want to change the terminology: free *will* may be a bit of a misnomer, but you still have free decision making, which is what we have meant for centuries by free will.

That you are hungry is not in your control. That you have learned to eat more at parties is conditioning. That you know cheesecake will make you fat or aggravate your diabetes or cholesterol is also learned; it is conditioning. That is why you are having that dialogue with yourself; you are well conditioned. Your decision *is* going to be based on that conditioning, especially if the choice must be made instantly (the tennis smash, the fastball). But another part of your brain's conditioning as you mature—infants do not do this instinctively, but you do it—is the realization that you are free not to like the options you come up with. You can try to think of other options. (Perhaps you will ask if you can take the second piece of cheesecake home for tomorrow or protest, "Stop tempting me! I love your cheesecake, but I just can't!") Yes, this process is going on unconsciously or preconsciously. Sometimes you'll just grab what you want. Because you have trained yourself over the years, though, and have been trained by parents and teachers of all sorts and by life's school of hard knocks to make your choices thoughtfully and intentionally and with regard to their outcome, learning, at least to some extent, to think critically, you remain responsible for your own conditioned decision making. Total or radical freedom you do not have. A great deal of freedom you do have. "Because you have trained yourself ..." is the key phrase here. *Of course we respond based on our conditioning. But we are, to a very significant degree, self-conditioned.* We would not even want "freedom" to mean deciding randomly, regardless of what we have learned over the years about what works and does not work in our lives, what is good for us and others and what is not. That would glorify ignorance, hardly the path to the nobility and responsibility we seek as we yearn for free will. "Let your misfortunes reprove you, / Let your afflictions rebuke you," we found Jeremiah saying earlier (Jeremiah 2:19), to which we should surely add: And let your successes embolden you, and your talents encourage you. The feeling that each significant decision is consciously made is an illusion. But the conditioning that operates before consciousness is not simply foisted upon us. We help program our own brains. We self-condition.

"But wait," I imagine some readers saying, "what if the temptation is a new one—no one ever tried to seduce me before or left a satchel full of money where I could take it and never be discovered. I have not been conditioned by others or had a chance to condition myself, to resist *all* temptations even if I wanted to do so." There is surely some truth to that, but not as much as we might naively think. We are not programmed to deal with *all* new possibilities. But we have read novels, gone to movies, and heard the gossip about others' peccadilloes. In both childhood and adult play we have had virtual experiences, identified with characters and imagined ourselves in their shoes. We have also learned paradigms—values systems, philosophies, religions, social mores—into which we may well place new opportunities in the process of deciding what is acceptable for us and what is not. At any given moment our freedom is limited by our conditioning. But we have in the past and continue in the present to have the opportunity to condition ourselves. ("I know there are going to be sinfully rich desserts at the party, and this time I am not going to have one!" "I know lots of people commit adultery, but that is not the sort of person I am willing to be!")

Consider an example in a different area. Indiana University neuroanatomist Jill Bolte Taylor explains that anger "is a programmed response that can be set off automatically. Once triggered, the chemical released by my brain surges through my body and I have a physiological experience. Within 90 seconds from the initial trigger, the chemical component of my anger has completely dissipated from my blood and my automatic response is over." From there on out, she insists, remaining angry or not is a choice. Especially once knowing this, we can learn—condition ourselves before the fact—to wait before striking out in anger. This is the traditional parental advice to the angry child to count to ten—though one would have to count very slowly to make that count last ninety seconds! The point remains, "Moment by moment, I make the choice to either hook into my neurocircuitry or move back into the present moment, allowing that reaction to melt away as fleeting physiology."[16] The anger hits, and is powerful, before we can stop it. But we are not totally helpless to deal with it. Learning that we can wait before acting in anger is one of many therapies, some of which we can master ourselves and some of which require professional help, available to help us program our brains. Just like the athlete who has practiced thousands of times how to return the tennis serve, we can and do practice and practice to develop our moves. For better or worse—thieves and murderers can do this, too—we program our brains to make us the people we choose to be.

If free will, then, is the ability to make a decision not at all based on instinct or preconditioning, conscious free will is an illusion, because decisions are made before we are conscious of them and will is an added feeling. But decision making as a complex process is very often, though not absolutely always, free. A conditioned process is not a predetermined process when the conditioning includes having learned to reject options our brains produce that do not fit our values and the paradigms from which they emerge and to search for more options. That, truly, is "free won't"! On rare occasion we may—we are free to—adjust or dramatically alter our values or paradigms. If we sense the need for such extreme reprogramming but still cannot come up with a way to do it on our own, we may seek others' advice or therapy. Not only our experiences but also our own thinking continue throughout our lives to influence our preconscious decision making. We are free not to do what first—or second or third—comes to mind. The decision-making process to which consciousness is a window is a free one. Because we self-condition, we have a great deal of what our traditions call free will.

BETTER AND BETTER, SO WHY NOT PERFECT?

Religious believers, and probably most secularists, too, are left with a paradox. On the one hand we insist that in any given case, because we have free will, we can do the right thing. If we do not like our choices, we can look for others. If we do not like our behavior, we can retrain, recondition, before the same situation recurs. Yet we insist that no one succeeds in doing the right thing all the time. I resist the temptation at this point to go into a theological exposition of different understandings of sin, from simple "missing the mark" (the root meaning of one of the biblical Hebrew words for sin, *chet*) to more abstract and psychological ideas of sin such as estrangement from God for having violated a divine commandment. Ritual matters, too, can wait. Simple but difficult choices illustrate the dynamic quite adequately. We need an explanation of why, if we have freedom not to sin, we sin anyway. From needlessly hurting others' feelings to sometimes lethal assault and countless violations of one another's dignity, rights, and property in between, we fall morally short. We should not evade the question by saying we do not always know what is right, for even when we know, we too often fail. If we are capable of making *each* decision correctly, we should, at least hypothetically, be able to make *all* decisions correctly. But we do not. With education, discipline, good intentions, and maturity, we can get better but cannot achieve perfection. There must be some flaw in us somewhere. What is it?

Education, we have seen, is a matter of conditioning, including self-conditioning. As we become better conditioned we can not only play better tennis, but we can also behave more ethically. The problem, though, is only partially in our conditioning. Our behavior is also circumscribed by our nature, the way our bodies and especially our minds work—our hardwiring, if you will. The body, for instance, needs food. We do not just want but need food. We perish without it. So from infancy—this is a genetic program—the brain monitors chemical balances and the sense of being physically "full." The infant feels hunger and signals it very effectively. He or she will later learn to postpone eating and further learn about healthy and unhealthy eating, sharing available food, even table manners, a raft of strategies for meeting the need for food in effective and socially acceptable ways. There is also hardwiring, here a metaphor for the evolved structure of the brain that provides us the capacity to learn all of this. But the hunger and the urge to satisfy it come first, and the rest kick in later. To carry on the wiring metaphor, it is as if there were primary circuits and secondary ones. Important though the latter are, and whether innate (cry and mother will come) or learned (smile while eating or compliment the cook to be given a larger portion next time), they exist to advance the primary. So, too, with the need for intimacy, which for adolescents and adults means, in part, a genetically programmed urge for sex, but is already present in infants, who, it has been convincingly demonstrated, do not merely like or want but *need* physical and emotional contact.[17] With age and experience we learn which behaviors will best lead to satisfying the primary needs, but the urges are primary and the strategies and self-control are secondary. So although our needs are all genetically provided for, the fact remains that as we preconsciously consider options, not all options are created equal.

The brain's task, we have said repeatedly, is to get each of us safely and successfully through life. Success in this context first means surviving, which requires food and other physical necessities, and ultimately reproducing. Translation (sticking with our examples): we are hardwired to want food and sex. We can learn, secondarily, that not all foods are equally good or bad for us and that not all sex is equally licit. But as our brains moment by moment generate options, not all choices are equal. When people get hungry enough, we will do most anything to obtain food. And when we are doing something important to us and a particularly sexy person comes by, that element of attraction will trump most other things we are doing at the time. Even if you are so self-disciplined that you would never—better say scarcely ever—consider acting on an illicit lustful impulse, you will almost certainly turn your head to see the scantily clad model (another sort of "cheesecake") amble by. To be blunt, we are hardwired to satisfy our desires—and not just for food and sex but also for wealth, pleasure, power, and honor. Anyone who has worked with children knows we are innately selfish. We want what we want, and we want it now. We learn self-control, sharing, respect for others, and the necessity of following rules in order to get along in groups. The selfish impulses come first; then we struggle to control them.

This all fits well, not surprisingly, with neo-Darwinian natural selection theories. Returning one last time to the cheesecake, as humans evolved those who liked high-fat and high-calorie foods were better adapted for survival and more likely to reproduce, therefore, than those who might have preferred vegetables. When food was plentiful—say the hunt was successful and the clan gathered to feast—those more inclined to overeat and store calories as fat for the time when the hunt was not so successful were also more likely to survive and reproduce than those with a modest appetite. Sweets, likewise, as sugar in honey or fructose in fruit, contain nutrients good for quick energy. We evolved a craving for fatty and sweet foods long before our modern era of sedentary occupations, affluence, and readily available food of all varieties. The adaptive urge that used to help us survive now predisposes us to obesity. As we self-condition, we tend to go on crash diets that help short-term results but do not get us over the urges. So the diet ends, and we tend to revert to earlier behavior. As for the other kind of cheesecake, the licit sexuality may well satisfy the craving as well as the illicit, but the urge, we each know, is very strong.

Fortunately, as we have seen, we also have hardwiring to check these and other self-serving impulses and censor our behavior. Still, as we create options and imagine consequences of our actions, some impulses are stronger than others. In a word, we are hardwired to be selfish. As we grow up, we learn often to delay gratification. But the urges remain—else, as the Rabbis said, no one would marry, build a house, or start a business, all physically and emotionally, as well as literally, expensive and risky undertakings. This is no more the body's fault than the mind's, especially because the mind, as a function of the brain, is embodied. If we separate out the mind-body dualism from the notion of original sin, an inherent tendency to selfishness remains. In biblical language, "the devisings of man's mind are evil from his youth" (Genesis 8:21). "Evil? Then their Creator is evil!" the Rabbis objected. All right, they reassured themselves, and us, these selfish inclinations are only bad if we let them go to extremes. Still, by our very nature, they are apt to be privileged when values conflict. We can learn and mature, conditioning ourselves to be at least a little better, but "we are not so arrogant and stiff-necked as to declare before You, our God and God of all ages, we are perfect and have not sinned; rather do we confess: we have gone astray, we have sinned, we have transgressed."[18] Freedom is not radical freedom. We cannot start over with a new brain or a wholly different personality, but neither are we helpless puppets of our past. Although the church was a little off on original sin (chuck the mind-body dualism), and the classic Rabbis a bit simplistic in their psychology (you are not an entity separate from your urges), they were both on the right track. We cannot be good all the time because we are hardwired for preferencing selfish inclinations.

There is no lack of guidance, traditional and modern, for dealing with this inherent flaw, and no lack of reinforcement—religious, philosophical, societal, and feelings of pride or shame as we do or do not control our impulses—aiding us to control and properly channel our impulses. When, inevitably, we fall short, religious traditions address our feelings of guilt, providing for confession, repentance, and renewal (we shall give some attention to this in chapter 7). Still, the lusts and greed are hardwired and primary, the controls, many of them at least, learned and secondary. To be otherwise would require having different brains. In personal-God language, that is simply the way God made us. More scientifically—and theologically if God is understood as inherent in nature—that is how we evolved. Our personal and collective struggles to be better are noble. Yet no one, not even God, can expect us to be perfect. We have free will and the potential dignity that is its corollary, yet no mortal makes *only* right decisions.

6

Morality
The Hop of Faith

Free will can only help us choose right and eschew wrong if right and wrong exist. Modern psychology provides evidence that certain values we would call moral are "hardwired" in our brains, and primatology and evolutionary biology suggest how that morality evolved. Because one of the ways in which God may be understood is as Order in the universe, it is arguable that just as animate being evolved out of inanimate being, so intelligence and then morality evolved out of lower forms of animate being. That there is a human as well as a divine element to our ethics should give us a degree of humility. No single teacher, book, or community has a monopoly on right answers. Yet to make life choices, not falling into relativism and indecision, we must act as if we know ultimate moral truth.

Once we learned one truth, and it was cherished or discarded, but it was one.

Now we are told that the world can be perceived by many truths; now, in the reality all of us encounter, some find lessons that others deny....

Yet we sense that some acts must be wrong for everyone, and that beyond the many half-truths is a single truth all of us may one day grasp....

Meditation from *Gates of Prayer*[1]

Morality is what many of us yearn to see realized in human society, the universalizing of right behavior. What we need is to see how humanity might have not only descended physically from earlier states of being but also ascended morally from them, ascended, that is, from primitive beginnings to positive and purposeful behavior that takes the welfare not only of ourselves but also of others into account. In Western religions we even dare to dream of a messianic future when peace and harmony will at last be realized. To achieve a moral order we must have confidence that there is such a thing as morality. A measure of freedom is vital, but there must be a good to choose, a bad to eschew. Whether for individuals or societies, only with ideals and goals to measure them against are choices meaningful. How, then, do we know right from wrong?

Science has taught moderns that we are descended from earlier life forms and, even before that, from inanimate energy and matter, which, step by step over eons, morphed into primitive life, then further evolved into animals and, very recently in evolutionary time, into people. Is there morality in this process? From the big bang through the highest forms of animal life, and human life is probably no exception, what lasts is what works, what somehow makes accidentally produced new forms better suited to endure and replicate themselves. Evolutionary biologists call this adaptation. That which survives is that which is adaptive, which gives a survival advantage. Evolution has no conscience. People, products of evolution, do. In that shift we should find clues to what morality is. We like to think of morality as our most noble achievement. We call "moral" what we believe is the right way to behave, whether seemingly adaptive or not. We like to bring love and human dignity, and just and compassionate treatment for other people, into the picture. Then we add to the realm of our moral concern decent and responsible treatment of animals and concern for the well-being of the biosphere and even of outer space (which we have begun to pollute).

We may argue that, in the long term, honesty is the best policy and that being charitable, compassionate, and even self-sacrificing bring us rewards as our families and society enjoy a more humane, civilized environment. But at the moment we face the difficult choices—whether or not to steal unguarded money, to remain faithful to a spouse, to feed the starving stranger, and so on—the moral path scarcely seems adaptive. That goes double, moreover, for those who take vows of chastity to serve God and church or who throw themselves on a grenade to save the lives of others. They will not pass along their genes. In such cases idealism appears the very opposite of enhancing the likelihood of progeny. This relates, of course, to the discussion of free will in the previous chapter. We are biologically programmed to take care of ourselves but also to look beyond ourselves to the needs of those closest to us, especially family (think of the maternal instinct for starters). Yet we appear most noble when we devote ourselves to the well-being of those beyond our immediate circle. Following basic rules of conduct is moral, and then going "above and beyond" the minimal rules, even to our own hurt, is nobly so. At the societal and planetary level our survival as a species appears dependent on humanity pursuing more than our selfish ends, balancing them with concern for the well-being of (at least) human others and then of the cosmic system of which we are a part.

We saw in the last chapter that we have a significant measure of free will, which is reassuring. Unselfish choices, though, do not come easily. Children must mature into their intellectual capacity over many years, and if abused in childhood they may never be fully free. Once a brain is badly programmed, it may be extraordinarily difficult to reprogram. Abused children are more likely to be abusive parents. Pedophiles are regarded as extraordinarily difficult if not, in many cases, impossible to rehabilitate. Even where the brain has not been so dangerously programmed early on, like other organs it is subject to disease and trauma—cancers, strokes, damage from concussion, dementia, and more. For all that, we still can—most of us, much of the time—choose between right and wrong. That brings us back to the philosophical conundrum. What is right? What is wrong? How do we know which is which?

We might say that there are certain principles about which everyone agrees: it is wrong to intentionally cause pain, for instance, or to kill others. Then your teenager says she is bored with her boyfriend but cannot think of a way to break up, for she knows how hurt he would be. That's different, you say. Some pain is unavoidable or even positive. A surgeon causes physical pain in the process of saving lives. Your daughter does her boyfriend harm, not good, remaining in an unhealthy relationship. Her own well-being is surely relevant, too! In Jewish tradition we quote Hillel, a sage from the first century BCE, who advised a balance between selfishness and selflessness: "If I am not for myself, who will be for me? But when I am only for myself, what am I? And if not now, when?" (*Mishnah Avot* 1:14). Do we at least agree that killing is always wrong? No! Someone comes after you with murderous intent, gun in hand, and you try to smash his skull with a nearby andiron before he shoots. Was that wrong? That's different, too; it is self-defense, we claim. Perhaps your defensive ploy fails and you are murdered. Now the state wants to execute the attacker for premeditated murder. Is that wrong? Opponents of capital punishment think so. Others disagree, and both point to sacred texts as they make their arguments.[2] We agonize similarly over the possibility of "just wars."[3] Without pretending to have the final answers to these and countless other such questions, surely we can agree that right and wrong are not always clear.

Attempting to evade the question, we could say that pragmatically, governments promulgate laws, representing societal consensus, and though they may vary in relatively minor ways, people should and do accept them. Americans remind ourselves that governments derive "their just powers from the consent of the governed." Consent? I was not there in 1776 when the Colonial Congress passed the Declaration that included that phrase and may well find this or that law, even in the Constitution, repugnant—as did opponents of slavery from the start, and advocates for women's rights more recently. Adolf Hitler was elected to office and set about limiting the rights of Jews, Gypsies, homosexuals, and others, eventually exterminating them. The majority of the populace, at the very least, acquiesced. By what standard can we say what he did was wrong? Societal consensus is important, but scarcely a guarantee of morality.

Philosophers from Aristotle to Mill, Kant, Rawls, and many more have offered moral philosophies, including meta-ethical principles such as basic human rights and greatest-good-for-the-greatest-number sorts of approaches, to provide a way out of this intellectual impasse.[4] Helpful though they are, the philosophical approaches disagree with one another, and none has been universally accepted.

Western religion has offered a different path. How do we know what we should do? God has told us! Or so Judaism, Christianity, and Islam have generally claimed. Through revelation, first to Moses and then to other prophets, God spelled it out for us. God, the Torah says, revealed *mitzvot*, "commandments," most notable among them the Ten Commandments at Sinai, at least six of which are ethical (honoring parents, and the prohibitions of murder, adultery, theft, testifying falsely, and coveting other people's property). The first four have ethical relevance as well, for loyalty to God and not following false gods should motivate obedience to the other commandments, swearing falsely often entails deception, and observing the Sabbath requires not only one's own cessation from work, but also allowing slaves, servants, and even animals a rest (Exodus 20:1–14; Deuteronomy 5:6–18). That is only a start. There are scores of ethical commandments among the 613 commandments that Jewish tradition counts in the Pentateuch, all of which, the text says again and again, come with divine authority. The same argument—doing right means obeying God's commandments, and doing wrong means disobeying—applies no less to ritual or other less purely ethical matters. It all worked very neatly—and not only in classic Judaism but in traditions rooted in Judaism, Christianity, and Islam, as well. Scriptural texts had to be interpreted to fit novel situations, but religious authorities found ways to do so. Vehemently though authorities argued over this or that application of the divine moral law, there was faith that we do have a divine law to apply.

Modern literary and historical criticism fatally undermined that consensus in the nineteenth century. First the Torah and then later biblical books were shown to be written by many human authors over many centuries, then edited, sometimes skillfully and sometimes clumsily, in later generations. Needless to say, there remains a tremendous amount to be admired in scripture, but we need no longer defend the indefensible, such as commandments for executing witches (Exodus 22:17) and Sabbath violators (Exodus 31:15), regulations for slavery (e.g., Exodus 21:2–6, 21:20–21, and 21:26–27), or instruction to commit genocide against the Canaanites (Deuteronomy 20:16–18). Nor (stepping outside the moral realm to emphasize the point) need we make apology for primitive understandings of creation (Genesis 1–2) or medicine (Leviticus 13) or for impossible suspensions of the natural order, like Joshua causing the sun to stand still (Joshua 10:12–14). We need no longer rationalize away contradictions. Did God create man and woman at the same instance or as two separate acts (Genesis 1:27 and 2:18–24)? Does God hold the children and grandchildren of sinners responsible for their ancestors' acts or only hold each individual responsible for his or her own deeds (Exodus 20:5 versus Eziekiel 18)? Now we simply say these composite texts are products of their time, clearly human, and therefore fallible, documents.[5]

The same sorts of historical and literary critical methods have been applied to the Christian Scripture, Qur'an, and Jewish "Oral Torah"—Talmud, midrash, and so on. Literalists in all camps have desperately argued that once human origins are acknowledged, we are on a slippery slope: soon we would not be sure any given verse, not even "You shall not murder" (Exodus 20:13; Deuteronomy 5:17) or "Love your neighbor as yourself" (Leviticus 19:18), could be guaranteed to be the literal word of God. They are right. Few doubt that there is a great deal of moral truth to be learned from these religious classics, but once even parts of them are acknowledged to be human, not divine, we reach an epistemological dead end: we must decide which is which, and then we, not God, become the decision makers, precisely what we sought to avoid by relying on revelation and scriptural authority. For better or worse, though, the literary evidence is overwhelming, however vehement the warnings against undermining faith in scripture may be. These texts convey insights worthy of a lifetime of study—and much aesthetic brilliance, too. But they are human words, not God's words. So although we can and should consult them to better understand many of the moral values of our civilization, they cannot serve as the sole foundation for morality. That is not to say, I must add, that God could not have had something to do with them. That is unknowable (except in the sense that if God is part and parcel of the order of the universe, then God has something to do with everything). Nor is there any reason to doubt that many of the authors believed they were faithfully conveying divine values. These books are sacred (*kadosh*, you will recall, means "set aside as special") because they are among the pillars of our faith, pointing us to God. Many of us sense we encounter God in these books.[6] Scripture remains for moderns a wellspring of faith, the work, very often, of God-intoxicated people struggling to apply the highest religious values to life. Yet for all our genuine admiration of religious classics, we cannot know right from wrong based on scripture *alone*.

That word "alone," I suggest, is more significant than it might at first appear. Societies certainly make good use of the idea of consensus and social contract, even if by themselves they are not a sufficient logical basis for morality and law. Ethical ideas from philosophy, political science, law, religion, and other disciplines serve as useful tools as we deal with the issues I raised above and innumerable others. When we must decide, for example, whether or not to pull the plug on a fancy piece of medical equipment that is keeping a patient alive, but with no real hope of recovery, I would not want to be dependent solely on texts from an era that never dreamed of electricity, much less of such medical technology. But neither would I want the decision made without the conviction that human life is in some sense sacred and not to be ended cavalierly or based primarily on financial criteria. To have more than one arrow in our ethical quiver is highly desirable as we make novel and agonizing choices. Our ideal, though, let us not forget, is greater certainty that we are doing the right thing—indeed, that there *are* right and wrong things. We have a gut feeling that right and wrong genuinely exist, seemingly without the ability to prove it. Cognitive studies, we shall see, prove helpful in identifying the sources of this vague sense—I would call it *faith* for going beyond the evidence—that right and wrong are real, not arbitrary choices. How nice it would be to find morality inherent in the natural order!

OF TROLLEY CARS AND OTHER MIND GAMES

Hypothetical cases involving runaway trolley cars, used by British philosopher and ethicist Philippa Foot[7] as far back as 1967 and subsequently a staple of theoretical discussions in many related fields, give insight into the nature of ethical decision making. Imagine you are standing by a trolley track leading down a hill toward you and, further, to a narrow bridge over a gorge. You see a trolley hurtling down the track and realize the conductor has fainted. On the bridge over the gorge are five people who have no time to escape the runaway trolley. But there is a sidetrack with one person on it who is unaware of the whole situation. If you throw the switch, the trolley will go onto the sidetrack and surely kill the one person, saving the five on the bridge. Is it morally permissible to throw the switch? Overwhelmingly, people asked this question respond that you are justified in throwing the switch, saving five innocents at the cost of one innocent.

Now imagine the same basic scene, except you are on a footbridge over the trolley track, and a very large stranger is sitting on the railing directly over the track. Unlike you, he is so large that if you give him a quick push so that he falls onto the track, the trolley will hit him and be stopped, but the large man will surely be killed. There are five people on the bridge over the gorge. Is it morally permissible to push the large man off the bridge? Overwhelmingly, people respond that you may not push him.

As a mathematical calculation, the examples are the same. May you cause the death of one to save five? So why do most people say yes in the first scenario and no in the second? If you ask them, many will refer to a religious or ethical principle they learned when growing up. Perhaps the "Golden Rule" will come to mind, and, because no one would relish being pushed off a bridge, "doing unto others what you would have them do unto you" means not pushing. But if you were the one on the sidetrack in the first scenario or one of the five on the bridge over the gorge in the second, would you not want someone to save you? Some people can scarcely answer the experimenter's question about why they chose as they did. They stammer incoherently or simply insist that saving the people in the first instance "is just right" and killing the individual in the second "is just wrong" Possibly the common reactions have to do with throwing a switch being less personal or emotional than pushing a person to his death.[8] Or maybe we instinctively feel that it is all right to cause harm as a by-product of a well-intentioned act, but wrong to use harm for that purpose.[9]

Brain monitoring does reveal more emotional activity in the second decision as people think of actively pushing someone to his death.[10] But more is probably involved, and the reasons behind the decision are less important than the fact that the experiment has been repeated in many places, and no one has found "evidence that gender, age, or national affiliation influenced the pattern of permissibility judgements."[11] The overwhelming consensus is not about the reasons, but about the answers. The all but inescapable conclusion is that our decisions in such instances come not from abstract ideas we have gained over the years, but from inborn values. We add the intellectual rationales after the decision. That there is near-universal agreement suggests powerfully that *our brains are preprogrammed with certain moral values*. In these cases those values are that under many circumstances some damage may be done to prevent worse damage, but not under *all* circumstances, for you may not actively kill someone even to prevent worse damage.

As we might expect, the extent of the consensus about these trolley scenarios changes if we alter the conditions. We are less likely to send the trolley onto the sidetrack if the one person there is a member of our family and are less hesitant to push if a large animal, not a person, is on the railing over the track. The multiplication of trolley scenarios and speculation about the underlying brain processes led a critic, Princeton philosophy professor Kwame Appiah, to dub them "Trolleyology."[12] We shall return to the critique, but be aware that there are plenty of other such hypothetical scenarios that also strongly suggest inborn moral values.[13] Jonathan Haidt, for instance, a psychologist at the University of Virginia, created the following scenario to test attitudes about incest. A sister and brother are traveling together in France on vacation from college and decide one night that it might be interesting and fun to make love. She is on birth control pills, and to be extra safe, he also uses contraception. They enjoy the experience and think the memory will always make them feel closer to one another. Still, they decide never to do so again but rather keep it as their special secret. The question, then, is, was this wrong? Nearly everyone has the same gut reaction: Of course—it is incest! Pressed for reasons, subjects respond with concerns about inbreeding or worries about the couple's future mental health. But there was ample birth control, they seem to have harmed no one else, and they are happy about the experience. "It is just wrong; I can't tell you why, but it is!" subjects insist.[14] As with the trolley examples, the explanations appear to be confabulations—the brain unconsciously and automatically making up a reason when it does not know why it had the reaction it did (recall Dr. Gazzaniga's split-brain patients confabulating back in chapter 1). Again, the brain is preprogrammed, hardwired, with certain moral values. With many such basic instincts, the conviction comes first, and our reasoning, including religious explanations like "God said so" and more philosophical or psychological explanations, come as after-the-fact rationalizations. The taboo evolved with our brains, presumably because incest was maladaptive and those who felt repulsed by it produced more offspring that those who did not.

Harvard professor of psychology, organismic and evolutionary biology, and biological anthropology Marc Hauser deals with all of this with considerable subtlety and thoroughness in his book *Moral Minds*. In addition to the hardwired values we have just seen demonstrated in the trolley and incest thought experiments, he presents further examples, including fairness, a universal concept lying beneath each culture's code of justice.[15] In an "ultimatum game," he explains, subjects are divided into pairs, and one member offers the other a share of a fixed amount of money, say ten dollars. If the second member accepts, they both keep their share; if he refuses, neither gets to keep any of the money. Most people offer five dollars to the partner. But some selfishly offer four or three dollars, and the offer is usually accepted, as you are better off with that amount than nothing. But when the offer is two or one dollar, the partner indignantly declines. Economically this is self-defeating; something is better than nothing. But there is an inborn sense of fairness, the flagrant violation of which triggers an emotional, not a rationally calculated, response. Most would rather have nothing—and the satisfaction of punishing the selfish partner! As Hauser summarizes the import of this, "The standard explanation for these results is that although we may have evolved as *Homo economicus*, we are also born with a deep sense of fairness, concerned with the well-being of others even when our actions take away from personal gain."[16]

No one yet has a full list of hardwired values or a complete catalog of which ones are strongest, though evidently certain ones trump others, as when you preference the taboo against actively harming the man on the footbridge over the impulse to save the five on the bridge over the gorge. Clearly, though, these hardwired moral principles are so fundamental to who we are that they pack a powerful emotional punch even before conscious thought about them, often compelling our action with no further consideration of options. We react instinctively: "Save those people on the bridge!" "No way I could sleep with my sister!" Only then do we step back to ponder why, and if we do go against the first instinct, we are apt to feel queasy or be left with lingering feelings of guilt or shame.

Recognizing such basic moral instincts does not mean that we do not *also* think many problems through or that cultural variation does not produce remarkably different moral codes from society to society. Hauser explains this as analogous to Noam Chomsky's linguistic theories, which hold that certain properties of language amount to a universal grammar, so that even widely differing languages share certain similarities, and our innate language faculty enables a child to learn any one of them. Likewise all moral systems share certain similarities, a grammar of the moral mind, as it were. Hauser writes, "We are born with abstract rules or principles, with nurture entering the picture to set the parameters and guide us toward the acquisition of moral systems."[17] Thus everyone with a normal brain has a hardwired aversion to killing people.[18] To lack that or other hardwired moral values is to be, by definition, a psychopath.[19] However, even for the normal-brained majority, as we saw at the beginning of the chapter with some people ready to make exceptions to the anti-killing instinct and others not (for war, abortion, or capital punishment), the specifics of how instinctual principles are applied vary. Likewise all societies have incest taboos, though they not uncommonly differ in detail, such as whether cousins may marry one another. In morality as in language, multiple sets of societal norms culturally evolve, with universal principles hidden beneath the diversity.

EVOLVED VALUES

Biologist and primatologist Frans de Waal, who teaches psychology at Emory University and directs the Living Links Center at the Yerkes National Primate Research Center in Atlanta, agrees that humans have hardwired moral values and takes the search for their origins back beyond human evolution to the evolution of prehuman species. He debunks the widespread belief, based, he says, on Thomas Huxley's misreading of Charles Darwin, that the natural world at all levels is an arena of pure selfishness, the eat-or-be-eaten jungle where, prior to the evolution of humans, nothing like morality is known, and even among our species' morality is but a thin veneer of civilization. There is a food chain and no lack of competition for resources both within and between species. But morality *also* evolved, Darwin believed and de Waal insists, because in many circumstances cooperation, reciprocity, empathy, and sympathy are adaptive. "It is fine to describe animals (and humans) as the product of evolutionary forces that promote self-interests so long as one realizes that this by no means precludes the evolution of altruistic and sympathetic tendencies."[20] Countless animals live in groups, the better to hunt, share food, care for young, tend to a lair, and so on. Many social animals appear to play with one another, to show affection for mates and offspring and loyalty to the group. Within the group, monkeys and apes constantly politic, giving favors and remembering who does and does not reciprocate, with males vying for power and mates and making alliances, and females similarly cultivating one another, gaining prestige within the group, and in alliance even calming or punishing aggressive males.[21]

When someone is upset or excited, and in response we become upset or excited, it is termed "emotional contagion." This is widespread not only among people but also among higher animals. Neurologically this empathetic response assumes at least some "theory of mind" (described in chapter 2), enabling a creature to think him- or herself into the position of another. Does the mother bird who feeds her chicks as they open their beaks imploringly understand and even empathize with their feeling of hunger? Probably not. She is probably following her instinct. But with monkeys and apes de Waal believes there is empathy:

> A rejected youngster may throw a screaming tantrum at its mother's feet, or a preferred associate may approach a food possessor to beg by means of sympathy-inducing facial expressions, vocalizations, and hand gestures. In other words, emotional and motivational states often manifest themselves in behavior specifically directed at a partner. The emotional effect on the other is not a by-product, therefore, but actively sought.[22]

Emotional contagion made possible maternal empathy, enabling mothers to know and care enough to act when their offspring were hungry, frightened, or otherwise needy. Empathy is not lacking in fathers, but women in general are more empathetic than men. De Waal explains:

> In mammals, parental care cannot be separated from lactation. During the 180 million years of mammalian evolution, females who responded to their offspring's needs outproduced those who were cold and distant.... The first signs of empathy—crying when another baby cries—is already more typical in girl babies than boy babies.[23]

This further evolved into family mutuality—reciprocated caring first for parents, then for siblings and others—and still further to concern among communities of animals (e.g., herds, packs, flocks) who depend on one another in many ways. Group as well as family bonding has survival value. Once the mental mechanisms developed for animals to empathize and sympathize, and thus defend and comfort one another, and the impulse evolved to reward reciprocators and punish non-reciprocators, these mental abilities could be adapted in other ways. Elephants as well as apes, for instance, have been known to demonstrate loyalty and emotional connection to the dead, seemingly mourning by returning to the dead body and even bones of mates.[24] Chimpanzees will comfort and help beaten or injured chimps, non-relatives, which does nothing for the passing on of their own genes.[25] De Waal explains:

> As so often happens [in evolution], the impulse became divorced from the consequences that shaped its evolution. This permitted its expression even when payoffs were unlikely, such as when strangers were beneficiaries. This brings animal altruism much closer to that of humans than usually thought....[26]

In book after book, de Waal provides anecdotes and formal studies, from his own research and the literature of his field, demonstrating what I would simply call animal caring, a combination of emotional and more cognitive concern about others.[27] Vampire bats who have a successful hunt, for instance, regurgitate blood to share with their less successful compatriots, first with offspring but sometimes with non-kin as well, and they remember who does or does not return the favor when the tables are turned.[28] Such reciprocality we see repeatedly in higher animals as well as in people. Higher animals are inclined to help others so others will, when the need arises, return the favor. In other words, the Golden Rule did not need people to evolve. Higher animals such as apes and dogs remember positive relationships, and abusive ones, too, despite years of separation. The stories we periodically hear, moreover, about cross-species sympathy and help, a mother gorilla rescuing a human child, a chimpanzee helping an injured bird, a dolphin giving a lifesaving ride to a shipwrecked sailor, are often based on fact, de Waal demonstrates, and clearly amount to help that is unlikely ever to be reciprocated. Once the help instinct develops, it can be generalized and need not be limited, in other words, to situations where there is hope of payback. As we just saw with emotional contagion and maternal instinct, what evolved for the preservation of genes is then available for unselfish use, too. This requires "caring about others and powerful 'gut-feelings' about right and wrong," de Waal insists, noting the theories of Antonio Damasio, which we touched upon in chapter 3, on emotion being crucial to motivate decisions and memory.[29]

Some would like to dismiss all alleged animal morality as, at most, enlightened self-interest and as unreflective (which is to say the mother chimp does not think, "Oh, I ought to help my child so she will grow up to be healthy," nor certainly, "... so she will pass on my genes," but simply reacts: "O, poor baby, let me help!"). But of course much human moral behavior is unreflective, as well. Even more to the point, de Waal is not arguing that animals are moral in as sophisticated a way as people can be, with our generally larger, more complex brains and with language to analyze situations and to produce moral philosophy. The point, rather, is that human morality, our generous features, if you will—love, self-sacrifice, compassion, and so on—are no less a product of evolution than our selfish features. "We're pre-programmed to reach out. Empathy is an automated response over which we have little control."[30] De Waal suggests the metaphor of Russian dolls, where the larger doll (us) has within it a smaller doll, which in turn has within it a smaller one, and so on. The higher moral endowments, like our physical endowments, are nested in lower levels.[31] Let us not forget, I would add, that moral and other intellectual abilities are more than simply *like* physical abilities; emotional and cognitive thought have a physical substrate, primarily the brain, that has evolved to make them possible. Morality, like all thought, is embodied.

In sum, Hauser and his colleagues make a plausible case for hardwired moral concepts in our brains, and de Waal and colleagues suggest how that hardwiring probably evolved. Earlier in the chapter, before turning to trolleys and evolution, I suggested how nice it would be if cognitive studies could show us that morality is inherent in the structure of nature. Philosopher Daniel Dennett, expounding on evolution, speaks of forced moves, "Good Tricks" that are necessary moves in the course of evolution. For example, to "fend off the gnawing effects of the Second Law of Thermodynamics," any living being is going to evolve a way of taking in new matter—we humans call this eating—or die.[32] You will not find a living thing that does not do this. They do not all eat as we do, quite obviously (think of plants), but they all somehow recharge themselves. Similarly, if we find creatures in outer space who have created a society, they will have evolved mathematics. They will not use our Arabic numerals, of course, but to share resources and build things they will have discovered basic mathematics (like $2 + 2 = 4$) and a way to communicate and thus use that information cooperatively.[33] With that as background, I invite you to imagine that a race of aliens arrives tomorrow from the depths of interstellar space. Depending on your tastes in science fiction, it does not matter for our purposes here whether they want to share information with us or exploit us for food. The question is, will they display, at least within their own culture, such qualities as cooperation, reciprocity, and—to motivate them—caring? Answer: of course they will! These are Good Tricks without which no culture could marshal the resources and knowledge to make intergalactic travel possible. Life could evolve without them, but not cultures. Whether on our planet or in the distant reaches of the universe, qualities and values we associate with morality are part of the structure of being. In theological language, if we think of God as the Creator of that structure, or in more philosophical terms as the Order itself, then these moral values are part of the Ground of Being and thus divine.

NOT SO FAST! THE NEED FOR A MORAL HOP

Good news and bad news. The good news: evolution has led inexorably to a level of sympathy, cooperation, even love, and to the dream, as yet unrealized but held up for us by religions and many a secular idealist as well, of full harmony among all. Likewise—negative values to go with the positive—there are universal aversions, among them murder, causing pain, lying, and cheating. Morality—not any single, perfect moral system, but the broad principles, the "grammar of morality" in our brains—is a product of evolution and of the very nature of being. *As life evolved out of inanimate being, intelligence and then morality evolved out of animate being.*

The bad news is that my qualifier above, "not any single, perfect moral system ...," is more than a niggling problem. First of all, even if the cognitive studies experts come up with a full list of the moral predilections that are hardwired in our brains (the science is young, but that seems a reasonable goal) and perhaps even a map of which ones trump others when values conflict, by what criterion could we say that what evolved is not only natural, but *right*? They may be what work best for us humans, but female praying mantises bite off the heads of their mates in the midst of intercourse, and some ant species, and chimpanzees at a far higher level of evolution, engage in aggressive warfare against other colonies. No one would call the insects or apes in these circumstances immoral. Morality requires understanding and making choices, and they are only responding to instinct. Our human instinctive choices, as we saw in the previous chapter, can often be modified by our rational abilities and our free decision-making capabilities. Not even we, though, can wholly ignore our hardwiring. Human morality, moreover, continues to evolve culturally. We have largely put polygamy and slavery behind us, but we continue waging war and debating torture and human rights. Who would argue that we have developed as far as we can or should? So even having found morality structured in nature, including human nature, how do we know that from a higher perspective—say a further evolved humanity or, in personal-God terms, in God's view—our human morality is yet fully moral?

Princeton moral philosopher Kwame Appiah observes that we may have evolved moral principles, heuristics, he calls them, that work in the most common circumstances we (or at least our ancestors on the East African savannahs) would likely face. Aggression being natural, to evolve reasonably peaceful societies we (and apes before us) needed to evolve an instinctive reluctance to hurting others "up close and personal." That explains why most people would divert the trolley in one instance but not push the man off the bridge in the other. When a snap judgment is needed,

> there is much to be said for an instinctive reluctance to push strangers over bridges, even if it might be, all things considered, in these horrific circumstances, the best thing to do.... That heuristic will lead us to miss the chance to save those five strangers on the trolley-track. But how often does a situation like that arise? As fast and frugal policies go, this one is very much more likely to lead us in the right direction.[34]

The persistence of an ethic evolved for a different age, I would add, has further ramifications today. In prehistoric times fighting required proximity, which may well explain why evolution made us less inclined to harm others "up close and personal." That leaves bombardiers and missile launchers far less troubled than perhaps they should be about killing hundreds or hundreds of thousands far from where they are pressing the button. Just because a moral principle evolved does not guarantee it is the highest imaginable principle in every case. We can take that as a challenge. Morality is real and important and is built into us. Still, we sometimes yield to lower instincts that are also part of us. We may aspire to do better at realizing morality in our lives and societies.

The moral principles are broad concepts and do not include unassailable criteria for applying them in every circumstance. As we saw at the beginning of the chapter, even after agreeing that murder is wrong—presumably one of the foundational moral principles—how do I know whether to enlist in a specific war, allow a pregnant rape victim to have an abortion, or inflict capital punishment on a murderer? Because of our language and abstracting abilities, we have come up with religions, legal systems, and moral philosophies. They all take our hardwired values into account but are culturally conditioned. To conflicting theoretical applications of the moral principles add an infinite variety of circumstances, and we are back to our need for decision-making guidance lest we be confused by the multiplicity of options. As we saw at the end of chapter 1, we need cultural paradigms, a set of credible beliefs, a structure of meaning, on which to base our lives.

This is important enough to linger over. One of the ways to conceptualize God is as the Structure of Being, and morality is part of that. Faith in a personal God may require a "leap of faith," but simply seeing order does not. We can recognize, and recognize that we are part of, a cosmic Order. We can see, moreover, that though there is tremendous wisdom in scripture, some of the morality there is primitive and that our morality, religious and secular, has continued to evolve. The leap that is required, and compared to the leap of faith this strikes me as just a hop, is that beyond the evolutionary logic of morality existing to promote human flourishing, *there must be a best way to construct human morality, though people may never fully agree what that way is.* Morality is real, though we have yet to find a way to make any given synthesis logically unassailable. To guide our lives we must rely on some specific cultural synthesis. We "hop" to Judaism or Christianity, to Kant or Rawls, to this or that legal system, or—most likely, consciously or not—to some personal eclectic blend of them, believing that in the process we can live largely moral lives. Lest we be paralyzed by uncertainty and indecision, we live *as if* our set of moral values captures absolute right and wrong, knowing, though, that as a human system, a cultural expression, it may at best come close to that ideal.

Are not all such leaps infinite? Some say so. But the very structure of our brains leads us to feel, passionately, that rights and wrongs exist. Experience injustice, on the negative side, or love, on the positive, and you cannot doubt the feeling, at the very least, that bad and good, and thus morality, are real. Neither is this pure emotionalism, for we rationally give assent to the idea that various moral principles make sense, though, as generations have put them together into systems, we cannot eliminate cultural variation. (We cannot with language, either, which scarcely means we cannot communicate at all even though we would communicate better if we all spoke the same language.)

There are, then, bedrock moral principles that have emerged out of the evolutionary process, and the evolutionary process is part of the Order of the universe. That the underlying values are divine but the systems are human suggests powerfully that the adherents of each religious system, Jewish, Christian, Buddhist, or whatever, and the proponents of different ways to organize societies, democratic or authoritarian, market capitalist or socialist, and so on, should take our systems very seriously, but not without humility. Humanity, "little less than divine" (Psalm 8:6), remains *less* than divine. We need our larger paradigms, structures of meaning that include morality.

But they are human products even as they encompass and may attempt to systematize divine values. Religions are not the only such paradigms, but they do serve that vital function for the majority of humankind.

Reiterating the question with which we began: are right and wrong real, that we might choose between them? Yes, but nevertheless the choice can be terribly difficult. There is guidance available aplenty, but no teacher, book, or community has a monopoly on truth.

7

Life Is with People
Organized Religion

The cerebral cortex, the higher memory area of the brain, evolved to help us keep track of groups. Higher animals, and then we humans, are naturally social creatures who not only cope with but also need community. Research shows that people do best in groups of approximately 150. Ironically, that is more people than many in our fragmented, mass society interact with in other than cursory fashion. Organized religion in general and gatherings for public worship in particular address this universal human need. Ritual, moreover, is ubiquitous in human culture, providing emotional as well as rational satisfaction. Prayer is more complex than simply talking to God—who may or may not literally "hear." Public worship, like art, expresses and evokes emotion, reinforcing identity and shared values.

*Cannot the heart in the midst of crowds feel
frightfully alone?*
Charles Lamb, in *Eliana*

133

W e were walking down the street in New York late on an August day in 2009. My wife, Ann, suggested we ought to buy a newspaper to check out upcoming events. We stopped at a newsstand, but they were out of the *New York Times*, then at a convenience store—also out. We tried a couple more places. "They're sold out because Michael Jackson died yesterday," my wife said. "That's ridiculous!" I responded, heading into another bodega. "Why doesn't anyone have any papers?" I asked. "Because of Michael Jackson," I was told. Ann was right. But so was I. It was ridiculous. The media had been saturated with news of the entertainer's death and continued to be for weeks after. Had the public not wanted to know every available detail of his last days and funeral plans, the views and feelings of his family, accusations against his physicians, and anything else they could learn, reporters and editors would have moved on to other topics. But there was tremendous interest. When I typed "Michael Jackson" into a search engine a month later, Yahoo! reported 1,280,000,000 sites where the name appeared. On a lark, I typed in "Moses": 91 million; "Barack Obama": 441 million. Could anyone register as high as the dead celebrity? I typed in "Jesus Christ," who was mentioned on only 179 million sites. Jackson's renown will fade. Six weeks later he was down to 818 million, and a year later to 658 million. With no disrespect intended to the memory of a talented entertainer, why in the world do millions of people who never met him want to know so much about Michael Jackson's life and death?

There is more here than the media's insatiable need to fill pages and screens. The computer news service I use as my home page offers me news of entertainers and athletes, and occasionally a scandal-plagued politician, at the top of the screen each time I go to the Internet. If weather or sports scores were more in demand, they would be there instead. Masses of people want to know celebrity news. Who was arrested again for drug possession? Who married or divorced? Who designed the starlet's ball gown? "Get a life!" I am tempted to say. Ironically, that comes close to explaining the phenomenon. Hold that thought. We shall return to it presently.

Animals that live in groups and cooperate need to keep track of who helps and who does not. Recall the vampire bats, which need blood to live. Not every bat finds prey every day, so as a matter of what is termed reciprocal altruism, an unsuccessful bat begs food from a successful one, who—Golden Rule ethic—regurgitates some blood for his companion, fully expecting that when the companion is the successful one in the future, the favor will be returned. He will remember whom he helped, and the hungry one will remember who helped and who refused. This ability to keep track of one another helps prevent cheaters from living off everyone's charity without helping in return. The human correlatives are easy to find, as when your neighbor refuses to loan you a hammer this week and then asks to borrow your ladder next week, or when a group is charged with a task, and some must work harder because of the slacking of others. Cooperation is crucial for many higher species, including ours. To limit cheating we need to recall who "plays fair" and who does not, so that the rewards are spread fairly each time or, when it is too late for that, so that at least the one who played you for a sucker this time cannot get away with it next time. Similarly, animals within their groups may compete for food or mates. They can be cruel, or they may comfort and groom one another. They join in political alliances to face down a predator or fight off a bully. For social systems, animal or human, to work, we need to remember to whom we owe favors and against whom we have grudges.

It takes a modest amount of memory to keep track of small groups, and more and more to keep track of larger and larger groups. Anthropologist Robin Dunbar found a direct relationship between the size of the brain's cerebral cortex (involved in higher cognitive functions such as memory, reasoning, and judgment) and the size of the typical group among various monkey and ape species. From the ratio of cortex size to group size, he extrapolated the size group that we humans evolved to prosper in, concluding that the human brain evolved to suit us for maintaining social networks of approximately 150 people.[1]

When we were evolving on the African savannah, we lived in smaller clans or tribes. In time we culturally evolved hierarchical social structures that enable leaders to keep track of reasonable numbers of lieutenants who monitor and direct further groups, and our language and mathematical abilities, plus technology (more and more rapid and long-distance communication, modern sanitation methods, computers), have enabled us to live in larger and larger groups. But even if you live in a city of millions you probably have far smaller groups of people whom you really get to know and care about, families and friendship groups not often much over 150.

In their book *Liars, Lovers, and Heroes: What the New Brain Science Reveals about How We Became Who We Are*, Steven Quartz and Terrance Sejnowski explain that while few Americans have that many close acquaintances, the number remains significant. Quartz, director of the Social Science Laboratory at the California Institute of Technology, and Sejnowski, director of the Computational Neurobiological Laboratory at the Salk Institute, note that businesses of up to 150 employees can operate without hierarchical management structures. "Once human subgroups become bigger than 150, they require hierarchical management structures." Military units are generally kept around that size so that soldiers will feel close and fight for their buddies at least as much as for the nation. The town hall meetings that impressed de Tocqueville in the 1830s rarely got much larger than that.[2]

Thus mentally equipped to deal with modest-sized groups, we are lonely and unhappy with long periods of isolation. The same brain power that probably evolved for protection against cheaters keeps track of benefactors and friends, the other side of the coin. We evolved to be social creatures. We *need* others. Our brains are structured to help us cope and flourish in modest-sized groups. Plop us down in the midst of thousands or millions of people and we feel anonymous and, ironically, alone. That 150 people or so is our "Goldilocks" number, not too big and not too small. We not only function well at that level but also, strangely, tend to be frustrated and unhappy well above or below it.

Quartz and Sejnowski suggest that in modern times cultural resources have had trouble keeping up with the development of mass society. Twenty percent of the American population lived in cities in 1870, fifty percent in 1920, and eighty percent today.[3] In smaller towns and cities people tended to know not only their immediate neighbors but also everyone in the neighborhood. Today many scarcely know the next-door neighbors. As is often observed, instead of sitting on the front porch and greeting passersby, we isolate ourselves in climate-controlled homes or apartments or relax on a fenced backyard patio where no one will disturb our solitude. A widely read book by Harvard's Robert Putnam, *Bowling Alone: The Collapse and Revival of American Community*,[4] tracked the decline of the groups that used to bring far more people together than they do today, from bowling leagues to political groups, book clubs, and, of course, religious groups. Membership in local chapters of national organizations rose from 1900 to 1960 but has steadily declined ever since.[5] Not only do the likes of the Elks Club, Hadassah, and PTAs have fewer members, but also on average people attend fewer meetings of the groups they do join.[6] In approximately the same period, the portion of the population attending weekly worship services has declined from 47 percent to 37 percent,[7] even though as high a percentage as ever report that they are religious believers. Writes Putnam, "Privatized religion may be morally compelling and psychically fulfilling, but it embodies less social capital."[8] "Social capital" is a key concept for Putnam, referring to "connections among individuals—social networks and the norms of reciprocity and trustworthiness that arise from them,"[9] or, in other words, exactly what not only the society in Putnam's view, but also individuals in Quartz and Sejnowski's, need.

Our society's mobility, I would add, scatters our extended families to the winds, leaving the nuclear family under tremendous pressure, with spouses challenged to meet all of one another's needs. When we go out other than to work, it is most often with spouse or children, *if* we have them—26 percent of the population lives alone. Alone or not, we return home each evening, tired from work, and station ourselves in front of the television in peaceful isolation. Hollywood's creativity is impressive, and we relax, repeating the process the next day, and the next. Then we wonder where the week went, or the year, and why we are somehow not as fulfilled as we would like to be.

In sum, we are happiest with 150 or so good friends, family, and close acquaintances with whom we interact regularly, a reasonable definition of a caring community. But that is no longer our norm. That brings me back to the celebrity phenomenon. The brain instinctively wants to know others, to keep track of them not only so they will be honest, but also out of a genuine desire to care and be cared about, to love and be loved. When we are starved for genuine community, we rejoice and grieve with the disembodied specters on our screens and in our earbuds. They are the people we know ... at least a little. So our restless minds adopt them. Fascination with celebrities, harmless in itself, is symptomatic of a deeper malaise in modern life, the lack, for many, of genuine community. There are other, less benign symptoms, at the extremes the psychopathology of loners who explode into violence, and in general the loneliness and alienation of mass society.

"DO NOT SEPARATE YOURSELF FROM THE COMMUNITY"

Providing a structure of meaning, a lens, as it were, through which to view the world and construe its meaning, is the prime function of religion. A secondary function, but far from a trivial one, is the bringing of people together in community. As we have just seen, this is a particularly urgent need in contemporary mass society. A community has been defined as a place where when you are absent you are missed and when you are present people know your name. Archaeological remains of early synagogues reveal that most would have met that standard, certainly in smaller towns, but also in cities, where they were often organized by a few families or a professional group.[10] With only a minyan (quorum for worship) of ten required for a worship service, the small size enabled anyone with the knowledge and ability to lead worship. The Talmud reports hundreds of synagogues in Jerusalem prior to the destruction of the Temple in 70 CE (*Megillah* 3:1 of Jerusalem Talmud; *Ketubot* 105a). Once Jerusalem and the Temple were destroyed, synagogues gradually became the center of Jewish worship, study, and community. Third- and fourth-century synagogues excavated in the Galil, the northern region of the Land of Israel, ranged from as little as 131 square yards, accommodating around 100 people, to the one that may be seen at Capernaum on the north shore of the Sea of Galilee, which measures 428 square yards and would have accommodated about 350 people.[11] The synagogue, then as now, was commonly called *beit knesset*, "house of assembly." Synagogues were not only for worship and study, but also for bringing fellow Jews together. Although a few very large synagogues are known to have existed around the Roman Empire—the one in Alexandria, for instance, where seats extended so far from the bimah that a scarf had to be waved so that people would know when it was time to say amen (Talmud, *Sukkah* 51b)—small was the norm, indeed small enough that we can well imagine all the participants knowing one another. Likewise, as Paul wandered from town to town around the Mediterranean Basin, the churches he visited were not physical buildings but groups of believers small enough to meet in homes. The apostle adjures the church members in Corinth not to quarrel (1 Corinthians 1:10–11) and to think of the church as one body in which no one part—eye or limb, and so forth—is superior to another (1 Corinthians 12:12–26). "If one member suffers, all suffer together; if one member is honored, all rejoice together" (1 Corinthians 12:26). In sum, creating community is not a new goal but a classic function of Western religious institutions.

For Jews as for Christians, facilitating community is not strictly a sociological function but one that is theologically and textually grounded. Hillel the Elder (late first century BCE to early first century CE) instructed, "Do not separate yourself from the community" (*Mishnah Avot* 2:4). As just noted, according to hallowed Jewish practice, worship services should not be held without a quorum of ten worshipers (only men in premodern times, either gender for liberal streams of Judaism today). Is the effect of prayer, whether on God or the individual, any different if it is recited by a group? Yes! In group worship, ideally, we sense that we are part of something larger than ourselves, a literal community first of all, and also an ongoing community that has been addressing God together, often in the same words, for centuries. I have been in congregations where a couple of thousand people prayed together and at conventions where many more thousands of people prayed, sang, and swayed to the music together. Most of those who conceive of God in personal terms, and certainly those of us who think of God more philosophically, would not say that the effect on God is any different. But there is an emotional component, too, to prayer. Worship affects us, and communal worship can be powerful.

People who move to a new area are often advised to go to church or synagogue, and clergy confirm that newcomers to our cities go "shul-shopping" and "church-shopping," seeking a place to "belong" and make friends at least as much as a place to pray. Modern American houses of worship offer refreshments and social hours before or after worship, an aspect of the total experience arguably as important as the worship itself. Denominational leaders and "church growth" literature and consultants warn that however wonderful the architecture, music, or sermon, if people are not greeted and made to feel welcome they may not be back at all, and certainly not regularly. This rabbi (and I claim no originality for this) rarely misses an opportunity to speak of the congregation as "the temple family," and like so many other congregations we have a Caring Congregation Committee to ensure that as members experience crises such as hospitalizations or bereavements, the congregation will reach out to them with concern, often tangibly in the form of a casserole. This sort of community building becomes harder and harder to do, or at least happens less and less spontaneously, as congregations grow in size from dozens to hundreds to thousands, making it less and less likely that the majority of members will know one another. Congregations generally try to address this by providing small-group entry points, groups for men, women, youth, seniors, empty nesters, and so on, as well as choirs, study groups, committees, and more. Finances for fancier buildings or more elaborate programming may come with size, but religion is not only about the more explicitly religious functions of worship and scripture study, but also about addressing people's needs. "All real living is meeting," wrote twentieth-century theologian Martin Buber.[12] That which touches the human spirit is spiritual, I suggested earlier. Only when we create community, helping to overcome loneliness and alienation, do we approach success as religious institutions.

Religious groups, quite obviously, are not the only place where community may be found. Yet they are exceptionally well suited to the task, first because our religions teach the importance of communal solidarity. Communal identity and religious solidarity are mutually reinforcing. A music group, professional association, or sports team builds on common interests as well, but lose interest in the violin, change career paths, or develop serious arthritis and you will almost certainly drift away from the group. Although the democratic ethos of American society implies mutual responsibility transcending socioeconomic, ethnic, and other differences, in fact neighborhood associations and many other voluntary groups tend to involve us only with people very much like ourselves. Groups based on commonalities less deeply rooted in our psyches may readily fall apart when a few disagreeable individuals become involved. Religious groups generally hold tighter and include a more economically diverse cross-section of people. (The diversity argument may apply more to minority religious communities, the one or two synagogues or mosques in a relatively large area, as opposed to the neighborhood Protestant churches seemingly on every corner in many cities. Still, some larger churches, certainly the so-called mega-churches but also many an urban church that develops a reputation for social justice concerns, youth programming, or what have you, do draw from diverse constituencies.)

Religious organizations, then, in our time as in antiquity, are an important source of the community we need. For all that, the era in American religious history when churches and synagogues built gymnasiums ended at least half a century ago, and though synagogues and churches may still run cultural programs such as film festivals, concerts, and lecture series, these are likely to stress religious identity or ethical themes, not just art-for-art's sake or knowledge of all varieties. As part of the "bowling alone" syndrome mentioned above, Jewish

Community Centers, YMCAs, and the like have been shrinking in number in recent years. A synagogue or church that fails to create community is ineffectual, but one that focuses on community to the neglect of more explicitly religious tasks of worship and religious education will likely be regarded as inauthentic. The two functions, fostering religious activity and creating community, depend upon one another. The community exists to put religious ideals into practice, and the religious ideals exist to enrich people's lives not only as individuals, but also in families, then in small groups, and ultimately and ideally in cities, nations, and the world at large. To put this differently, although the sociological contributions of religious institutions are significant, they only scratch the surface of the profundities carried by our ancient and ongoing religious traditions. Religions must promote faith and understanding, pointing us toward meaning in life, to serve their legitimate, important function as perspectives on life, higher paradigms. So we must not only gather, but also do so for religious exercises regularly. We must worship, study our sacred scriptures, and attempt to apply our values in society.

Perhaps nothing that religious organizations could do would immunize them against the forces of atomization and privatism that draw so many away from commitment outside their immediate families and occupations. Still, examples of religious groups that do well attracting members are available in most cities, and I dare say the common denominator is their success at making people feel welcome and needed. This is as important as the need to articulate theologies that educated moderns find compelling.

RITUAL AS ART FORM

Multifaceted though the programs of churches, synagogues, and mosques are, at their core we expect to find worship services, holiday observances, and life-cycle ceremonies. In a word, we expect to find religious *ritual*. Communal resources are lavished on buildings, clergy, and all the support services to facilitate this ostensible reaching out to God. Our religions insist that people can pray anywhere, and most agree that God is not so vain as to need our constant thanks. So, aside from the social benefits we have just been discussing, why all this expenditure of time, effort, and money on communal religious ritual?

Ritual is ubiquitous in human culture (and in much of the animal world, too, not so incidentally, as bees do their dances to lead others to a food source, social mammals pay obeisance to alpha males, and so on). Even before we get to specifically religious ritual, we should recognize that ritual must have evolved to meet some deep need or needs. Rituals are repeated actions with commonly agreed-upon meanings. In chapter 1, we noted that routine and habitual behaviors save us from constantly renegotiating common situations in which we find ourselves. Thus the social ritual of saying "Hello," "Goodbye," "How are you?" "Have a good day," and such are no more or less than the establishing and maintaining of a cordial social bond. You do not have to stop to think what you ought to say to show your friendly intent or take the time to find out what sort of day a stranger, casual acquaintance, or even a friend is having when one or both of you is busily engaged in other business. The meaning is in the act of wishing for or asking about the other's well-being. There is certainly an element of that in public religious ritual, as well. The meaning is at least partly in the showing up ("I'm part of this community ...") and in saying things that relate us to God ("... and I follow God's and the tradition's ways"). This is a message to others as well as to oneself, so even those who do not worship regularly, but show up for a couple of major holidays each year, are expressing identity, however strong or weak their faith.

Secular communal rituals often come close to the deep resonance of communal religious rituals, expressing and creating communal identity while expressing values. Participants at a political rally, with flags on display, special songs sung, and inspirational words offered, or at a football game as the team runs onto the field, the crowd cheers, and acolytes known as cheerleaders whip up enthusiasm, feel the almost magical excitement of the moment. Emotion is generated to support the cause, in the former case political and in the latter athletic. Group cohesion and enthusiasm are enhanced by rhythmic group chants ("Four more years!" or "We want a touchdown!") and familiar music.

Whether relatively trivial ("Good morning"), somewhat significant (saluting your superior officer), or potentially profound (pledging allegiance "to the flag ... and to the republic for which it stands"), rituals overtly or subtly remind us of who we are and how we fit into the social matrix. They help create and certainly reinforce values and loyalties. Religious rituals do all this, too, but extend the social matrix beyond our immediate social and political context to our place in the cosmos. This is all enacted like drama and elaborated in multiple aesthetic dimensions—beautiful language, music, art, and architecture. The ideas and ideals of the religious tradition are acted out, and not only verbally but also with all the senses. In effect, we speak and sing our faith, walk or dance it, touch, eat, and smell it.

Recall, moreover, that the arts both express and are intended to evoke emotion. Recall further that memory is greatly strengthened when enhanced by emotion. Why keep saying the same things time and again to God—whom some of us do not regard as literally listening—even if God does hear? There must be something deeply satisfying in religious ritual or it would die out. There is comfort in knowing where we stand in life and the world and in reinforcing the religious paradigm that guides so much of our decision making. But all that sounds very cerebral, and worship is not for the most part study, and certainly not systematic theology. Worship and other religious ritual, such as life-cycle ceremonies, appeal emotionally, in the process reinforcing cognitive content.

Identity, sociability, belonging, and values, enhanced aesthetically to evoke emotion, which reinforces it all—that is religious ritual. It appeals to cognitive areas of the brain, too, but the emotional satisfaction is what brings people back. A philosopher and atheist reputedly observed that if he were to practice a religion he would be a Roman Catholic. Why? his students asked. "Because it's pretty."[13] At the risk of trivializing, that is it. No need to be snobs about which forms our art should take. Some prefer classical-style music, and some folk-rock. Some thrill to soaring cathedrals, and some would rather sit in a circle face-to-face with fellow worshipers. For all the variety, people like public ritual first for the sense of belonging and then for the reasons we like other art forms. As important but no more important, ritual inculcates and reinforces the religious paradigm that has become a key part of who each worshiper is. That cognitive religious content matters. But we misunderstand the experience if the explicit content is allowed to eclipse the emotional resonances. Indeed, remember that spirituality occurs when an emotional surge reaches conscious level while you are thinking religiously related thoughts. Thus, if the ritual performance is effective, even the rational elements may move us emotionally.

WHEN PUBLIC PRAYER WORKS

What makes a group religious experience effective? Despite great variety, in general the aesthetics of worship derive from the language, often poetic, of prayer and how sensitively it is read. Then there is the music—choral or solo, passively heard or actively participated in—and the beauty of architecture and symbolic objects. The choreography of how worshipers and officiants relate to each other, as well as their standing, sitting, kneeling, swaying, and marching before God, contributes. In any given case there may be more, certainly the quality of preaching—and also whether or not the air conditioning is functioning! That final point brings us to an obvious further factor. As in any aesthetic activity, the effectiveness depends not only on the skill of those presenting, but also on the preparation and mood of the audience. A congregation of non-Catholics at a nuptial mass may well find it interesting or even somewhat moving but will not likely respond as readily or deeply as knowledgeable Catholics would, attuned as they are to the symbolism. Even when participating in the public worship of our own religious community, if we are exhausted, distracted by rude people chatting nearby, or preoccupied with other concerns and thus unable to give full attention to the worship, we will likely get less than we otherwise would from the experience.[14]

Consider, in all of this, that other explanations of religious ritual add further nuances to our understanding but miss the forest for the trees to the extent that they focus on one or another element to the exclusion of others. Clifford Geertz's influential essays "Religion as a Cultural System" and "Ethos, World View and the Analysis of Sacred Symbols,"[15] which highlight how prayers, symbols, and ritual point beyond themselves to a religion's reading of the world, shaping people's experience while expressing the religious group's ethos and worldview, are helpful. So is the new study *Ritual and Its Consequences: An Essay on the Limits of Sincerity*,[16] which speaks of the "subjunctive" function of ritual. Praying for peace in a community that believes a messianic era of peace will one day come, for instance, creates an "as if" world and helps resolve the tension between the unredeemed world of now and the redeemed world yearned for and promised by the tradition.

A more specifically neuroscientific approach comes from Drs. Andrew Newberg and Eugene d'Aquili (the University of Pennsylvania scientists who took pictures of brains as people prayed) in *Why God Won't Go Away*. They argue that repetitive ritual actions—rhythmic singing or dancing, for instance—may calm or excite certain areas of the brain. I believe they overemphasize the importance of rhythm and mystical states (plenty of rituals involve neither), but they are surely correct that "the emotional effects of ritual are dependent upon other body sensory input and, most important, the cognitive context in which religious ritual is performed."[17] In sum, worship is so widespread because it brings ideas alive aesthetically, moving us emotionally.

Few people will be regular participants in any religion's worship unless they have adopted many of its teachings as their own. But because religious ritual involves us more emotionally than purely rationally, and because most major religious traditions are very broad and there is thus rarely unanimity on doctrine, a variety of beliefs often coexist within worshiping communities without preventing people from praying together. The personal-God worshiper has obvious reason to talk to the "Sovereign of Existence" (*Melech ha'olam*), who is presumed to be listening. What about those of us who affirm one or another philosophical God concept—in my case, God as the Order of Being? The meditators' mystical sense of oneness with all being or the philosophical believers' urge to understand how we fit in and what we ought to do to make ourselves and our world better can also find satisfaction in communal solidarity and emotional stimulation. This is about our mental state, not God's! Pondering our beliefs, with a further boost from all the aesthetic resources marshaled for public worship, we, too, may well have spiritual moments.

Another example is germane. A review of the various types of prayers that religions employ would take us too far afield, but we should note in passing that little if anything in the secular world comes close to the power of religion where atonement is concerned. By appealing to God for forgiveness (personal-God image) or focusing attention on God to realign one's life with one's highest ideals (philosophical-God image), and doing so in the context of relatively elaborate rituals that hold out the promise of somehow cleansing the spirit, thus adding a crucial emotional element to cement the memory, believers feel renewed and thus enabled to move on in life with less burden of guilt. A prison sentence is more apt to leave the person feeling humiliated, and therapy requires a major commitment of time and expense with no guarantee of success. What was referred to above as the subjunctive nature of ritual, psychologically dissolving the boundary between the literal world and the ideal world, leaves worshipers ready to put past failures behind them and thus, in most cases, better able to move forward positively. A sense of God's reality, however personal or abstract, is needed for this to work.

Twentieth-century Jewish theologian Mordecai Kaplan, who conceived of God as a Process making for salvation or a Process making for righteousness, not a conscious being, wrote:

> Prayer aims at deriving, from the Process that constitutes God, the power that would strengthen the forces and relationships by which we fulfill ourselves as persons. We cannot help being aware of our dependence upon the Process which we identify as God, namely, on all that makes for goodness, truth and beauty in the world, for our success in achieving a mature, effective and well adjusted personality, and we naturally articulate that need in prayer.[18]

That is on target, and aesthetics are also involved and thus emotional appeal. Spirituality is evoked. Passions may be aroused. In worship we recommit ourselves and our communities to living our lives in consonance with the Reality within which we live.

Before digital technology, to get a clear radio signal or television picture we used to turn a knob until the static disappeared and the signal came in clearly. In that sense, I like to say, prayer is "getting on God's wavelength." Or, switching to a visual meta- phor, prayer is about bringing life into focus. Life's beauties and pains, and its yearnings, fears, and hopes, are better seen and coped with when both our own lives and the Divine come into clearer focus, enabling us, as individuals, participants in historical and contemporary communities, and members of the human family, to sense our place in the cosmic scheme of things. [19]

8

Why My Religion? What of Yours?

Why should we be loyal to one faith if, in fact, all the major religious traditions deal effectively with universal human needs? By the nature of religious paradigms and how they prevent our being overwhelmed, we need a firm commitment. Thus conversion between major religious traditions is the exception rather than the rule. Yet there is no one-size-fits-all religion. Within as well as between the various religious traditions, diversity is the norm. Stories appeal to us because personal identity derives from the way our brains make narratives of our experience. Thus our scriptures, broadly speaking our stories, unite us within traditions as we come to regard the communal story as our story.

> *When the Baal Shem had a difficult task before him,*
> *he would go to a certain place in the woods, light*
> *a fire and meditate in prayer—and what he had set*
> *out to perform was done. When a generation later*

the "Maggid" of Meseritz was faced with the same
task he would go to the same place in the woods and
say: We can no longer light the fire, but we can still
speak the prayers—and what he wanted done became
reality. Again a generation later Rabbi Moshe Leib
of Sassov had to perform the task. And he too went
into the woods and said: We can no longer light a fire,
nor do we know the secret meditations belonging to
the prayer, but we do know the place in the woods to
which it all belongs—and that must be sufficient; and
sufficient it was. But when another generation had
passed and Rabbi Israel of Rishin was called upon to
perform the task, he sat down on his golden chair in
his castle and said: We cannot light the fire, we cannot
speak the prayers, we do not know the place, but we
can tell the story of how it was done. And, the story-
teller adds, the story which he had told had the same
effect as the actions of the other three.[1]

Much of what we have examined involves universals: the way our brains create our world; the need for "higher mental paradigms" or "structures of meaning" through which to interpret our lives; the tendency to personalize natural feelings of awe; the emotional component that lifts rational thinking to spiritual experience; the measure of free decision making we enjoy and the hardwired moral values that religious and other ethical systems encompass; and the need for community. Our religions work the way they do because our brains work the way they do. If, from this cognitive standpoint, all religions work similarly, why, aside from accident of birth and perhaps inertia, remain loyal to one? Will any religion, then, do? Are they all equally "true" or at least equally efficacious for guiding lives and societies?

To approach that issue, consider where most of us get our religions in the first place. Human infants are born before their brains are fully mature, else their heads would be too large to fit through the birth canal. We are born with some basic biological talents such as breathing. But far more of the knowledge we will need to make our way in the world and to appreciate what life is all about requires years of experience and learning. Perhaps trusting your mother is hardwired, as with ducklings who trustingly follow the first thing they see move after they hatch,[2] or perhaps that trust is learned by the helpless infant in the early hours and days of life when mother, providing all the baby's needs, becomes supremely worthy of trust. Clearly parents especially, and then other authority figures, lay the foundations, as it were, for our religious life as for so much else. When a parent leads bedtime prayers or models other ritual behavior and early on sends the child for religious instruction, the scaffolding on which religious consciousness will rest begins to form. Because one of the chief functions of religious paradigms is to provide meanings and another is to filter out other interpretations, we should not be surprised that so many stick with the religion into which they were first acculturated. As the Jesuits reputedly said in the sixteenth century, "Give me the child for seven years and I will give you the man."

Contemporary religious philosopher John Hick devotes nearly half his book *The New Frontier of Religion and Science: Religious Experience, Neuroscience and the Transcendent* to the implications of religious diversity.[3] Hick, who has written much on comparative religion, notes that worldwide, people overwhelmingly are shaped by the religion of the family into which they are born.[4] Even in the multicultural environments of modern Europe and America, where "more people than in the past are acutely aware of faiths other than their own," most people do not ask which religion is best, "because they think that they know the answer already."[5] Hick presents two criteria an objective observer would need to regard a religion as successful at all, much less better than another. First, religions must deliver meaning to individuals. Hick calls this salvation, acknowledging that it is a Christian term but trying to employ it more generically as "helping people to find a limitlessly better existence, to which it shows the way."[6] All the major religions, he acknowledges, succeed at that for millions of followers. Second, religions attempt to motivate moral behavior in their adherents. Hick thinks they all do this, too, and all show themselves capable of producing outstanding and even saintly individuals. Religious cultures have their ups, downs, and periods of greater creativity over the centuries, but

> when we try to look more specifically at any faith's production of human goodness in ordinary people, and also of outstanding saintly individuals, it would be hazardous in the extreme for any of the great faiths to claim that its adherents are, morally and spiritually, better human beings than the rest of the human race. We have no statistics here, but certainly the onus of proof, or of argument, lies upon anyone who ventures to make such a claim.[7]

Thus we need a "theology of pluralism," and Hick contents himself with arguing for the reasonableness of such a theology. Borrowing a concept from Immanuel Kant, one very much consonant with cognitive studies that, as we have seen, speaks of how we create our reality, Hick argues that we know reality, including whatever of the Divine is part of reality, only through our own senses and cognitive structures.[8] We do not know any aspect of reality "in itself," including scripture or God. God "in Himself" or "in Itself" is ultimately unknowable. We know the image of God—and other religious ideas—that each of us has learned from a religious tradition. So if, once again, each of the religious traditions gets us to an image of God and a religious approach that does the salvific and morality-motivating job for us, and we know that other people, who also do not know God-in-Himself, achieve that goal, too, then we can acknowledge that more than one religion is legitimate. Like the classic blind men examining the elephant, our very different concepts may well be reactions to the same underlying divine reality.

That is fine as far as it goes, and a theology of pluralism is, indeed, a must for our multicultural societies and world. To my mind, though, that leaves us over the abyss of relativism, affirming at most that it does not matter what you believe as long as you believe *some*thing or maybe that it does not matter if you believe in a religious paradigm at all as long as you find something that works to help you find meaning and live morally. Have those who died rather than forsake their faith, then, been fools? Had they converted rather than being martyred, would they simply have been swapping one efficacious religion for another?

We need to maintain Hick's pragmatic pluralism while adding criteria for passionately affirming a single religion. Granted that, epistemologically, we cannot *prove* one path better than another, that does not mean that one may not, in fact, *be* better than another. We can and do know, after all, that the teachings of the various religions, though they overlap substantially, especially where ethics are concerned, are not the same. Even within each of the major religious traditions there are dramatically different interpretations, mystical versus rational, for instance, more and less fundamentalist approaches to scripture, and so on—the substantive differences that lead to denominationalism in its various guises. I am not even prepared to grant that, say (within my Jewish tradition), Orthodox, Hasidic, Conservative, Reconstructionist, Renewal, and

Reform Judaism are all equally valid in terms of truth claims and ritual requirements. All capable of satisfying various individuals, yes. And that is saying a lot. As a committed pluralist I am happy to coexist and cooperate as long as I need not affirm something that violates my conscience (e.g., that homosexuals are more sinners than anyone else, for instance, or that women should not be allowed to become clergy). In modernity, with appeal to the authority of texts undermined, I acknowledge that I cannot prove my liberal Judaism is better than Orthodox Judaism, and the Orthodox cannot prove the contrary. Or at least they cannot prove the contrary other than by denying my premise that the authority of texts has been compromised by historical studies. Denying one another's premises is precisely what we do. Strange as it may seem, that is a good thing, for in the vehemence of our conviction that "we" are doing it right and "they" are doing it wrong, each of us works harder to do our own thing *well*. A wishy-washy "whatever you feel like" religion will not maximize anyone's sense of meaningful living or motivate exemplary—"above and beyond," self-sacrificial, *holy*—behavior. We are each better off believing, not blindly, but based on our thoughtful consideration of texts and experience, that we are "doing it right" and the others' priorities are askew. That we cannot prove the other wrong does not establish that the other is right. If God were literally to speak, God would probably indicate that neither of us has things perfectly. Unable to prove one way better, we should acknowledge that the others' ways work for them as ours works for us. Yet that is only a pragmatic judgment and does not logically require that both are equally valid. This should engender humility. We have more scriptures, beliefs, and practices in common than not, however dramatically different our interpretations or the aesthetic styles of our worship.[9] We should each recognize that we might be wrong on any minor point or even, at least in theory, on a major point!

If even within each religious tradition we can disagree passionately, certainly between them, where the intellectual gulfs are substantially larger, this must be the case. Although there is an important grain of truth in the folk wisdom that all religions are different paths to the same God, that papers over profound differences in the way God is conceived and the nature of the paths. We could say that both Judaism and Christianity speak of salvation. But "saved" from what? From death, from enemies, from sin, from meaninglessness? And how and when might we achieve this state? It is hardly a mere difference in emphasis that most Jewish theology stresses group salvation, thus postponing it to the indefinite messianic future when we shall "perfect the world under the sovereignty of God, and all flesh will call upon Your name"[10] (*I* cannot be saved until *we* are), and most Christianity theologies see salvation as individual and available to people whenever they come to faith (the messianic kingdom is already here; "Christ died for *my* sins"). A chapter or book could be written only on salvation; the point is that even slightly expanding such common simplifications destroys the seeming sameness. The major religions may, indeed, as Hick tells us, all "work" to help their adherents make their way through the world—that, as I have argued throughout this study, is what religious paradigms do. But they are not all the same, nor, when within or between traditions there are flat-out contradictions, could they all be close enough to truth that it scarcely matters which one a lover of truth affirms.

In other words, Hick's purely pragmatic criteria—"By their fruits you will know them," he likes to quote[11] (Matthew 7:16, 7:20)—are important and reasonable, but inadequate. For a religion to become or remain a firm foundation for a life, there should be deeper reasons for loyalty than accident of birth. So let us delve once again into an aspect of the meaning-constructing nature of our brains, where, I suggest, further criteria may be found.

NARRATIVE, MEANING, AND MYTH: WHAT OUR STORIES DO FOR US

One of the vital things that our brains do is turn our experience into narrative, creating what we sense as a "self" in the process. I am the one who grew up in such-and-such a place, decided to do this or that for a living, had this trauma and that satisfaction, and made these choices. Currently I am doing what I am doing (me writing; you reading) and hope later in the day to do various other tasks. Longer term I have dreams and goals. In sum, I sense myself living my story. Someone who does not remember his past, generally due to a brain trauma or lesion, no longer knows who he is. Even with full awareness of what is going on around me right now, without knowing my past I would not know who I am. We know this as tragic in relatively rare cases of amnesia and frighteningly common cases of Alzheimer's disease and other forms of dementia. As we live, we make of our experience more than a story that we then believe about ourselves. The story in large measure constitutes the self. As philosopher Daniel Dennett puts it:

> Our fundamental tactic of self-protection, self-control, and self-definition is not spinning webs or building dams, but telling stories, and more particularly concocting and controlling the story we tell others—and ourselves—about who we are. And just as spiders don't have to think, consciously and deliberately, about how to spin their webs, and just as beavers, unlike professional human engineers, do not consciously and deliberately plan the structures they build, we (unlike *professional* human story tellers) do not consciously and deliberately figure out what narratives to tell and how to tell them. Our tales are spun, but for the most part we don't spin them; they spin us. Our human consciousness, and our narrative selfhood, is their product, not their source.[12]

No wonder you are so interested in stories; you are one! That is, you do not just have a history, but your sense of who you are is bound up in it. As you ponder your own story and that of others, moreover, you infer that earlier events cause or at least influence later ones. Further, you want to remember how you managed a situation where your decisions led to a positive outcome (e.g., you were fed, honored, mated with) so you can do it again, and you have at least as great a need to remember what did not work (ouch—don't do *that* again). As we are influenced by our stories, so, we come to realize, are others influenced by theirs. So if we want to predict what those around us are going to do—hug us or hit us, for instance—we want to know their stories, what is motivating their actions. Indeed, we are so naturally curious that we love stories, even fictional ones, from which, very significantly, we can also learn (if ever in a like situation, I should not make the mistake that the character in a given tale made, or should display that character's courage, and so on). Stories, history, even gossip, fascinate us because our brains evolved to interpret them. As consumers of food we do not have to think about feeling satisfied when hunger and thirst are slaked, and as consumers of information we do not have to think about feeling satisfied when we figure out plots, motivations, and likely next steps. We enjoy doing what we have evolved to do.

Furthermore—an observation that by this point should not surprise anyone—the intellectual aspect of the appeal of stories is accompanied by an emotional aspect. As we experience a story unfolding, true or fictional, we are naturally empathetic. We relate to and identify with others' stories, thus adding what happens to others to our store of personal experience. Recall the work of de Waal, discussed in chapter 6. A side effect of the evolution of empathy to nurture maternal, familial, and group concern and help was its availability for other applications. Here is an adaptive value for that nonfamilial and non-"in-group" empathy. We often share the characters' emotions, rendering even vicarious experience an important part of our experiential base.

Religions build on this aspect of our consciousness, too, communicating values and professed truths in story form. In the beginning, there is story. Abstract ideas, systematic theologies, catechisms, and doxologies come later. The Hebrew Bible starts with a story of creation, followed by story after story, and then, in addition, by commandments, histories, prayers, and proverbs (much of which contain stories, too). Christian Scripture begins with the gospel story, four times over. All religions, and thus most religious scriptures, are full of stories, and stories, implicitly or explicitly, carry values—attitudes, judgments, perspectives—that those who hear them may add to their store of experience. Stories widely shared by a religious culture are that group's *myths*, pointing beyond themselves to deeper lessons and worldviews, and ultimately to God. The God pointed to, moreover, is not God as defined by abstract propositions, but God as experienced by those telling the stories. To the extent that the subjective feeling of the perceived divine encounter is influenced by prior religious learning, this is the spirituality, an emotional as well as a rational phenomenon, the particular religious tradition offers.

Even within and certainly between religious communities, we may not agree on the historicity of any given story. But the lessons carried by, say, the Exodus story, or Jesus's birth and death narratives, or the account of the Buddha achieving enlightenment beneath the Bo tree, whether historical or only heuristic, become the collective memory, as it were, of the religious community that repeats them generation after generation. When we speak of religions as higher paradigms that shape our perceptions and our sense of meaning as each of our brains creates our world, the stories are at least as important as the abstract concepts, commandments, or rituals. The more we become acculturated to a specific religious tradition, the more we are as likely to see the world through the lens of its stories as through the lens of our personal stories. Indeed, as we hear stories repeatedly, reacting emotionally as well as rationally, the special stories of our religion and culture become our stories, too.

Religious traditions, in fact, often urge the faithful to regard the collective story as our personal story. This is literally absurd but symbolically profound. When a Christian preacher tells the faithful that sinful humanity crucified Christ and that God's self-sacrificial death redeemed them from sin, the preacher is making explicit the implicit message of the myth—in the high sense of the word "myth" as carrying a culture's values. The same holds when Jews attend a Passover seder, the ritual meal where the story of the Exodus is retold, and recite the following:

> In each and every generation people must regard themselves as though they personally left Egypt, as it says, "Tell your child on that very day: 'This is what Adonai did for *me* when *I* left Egypt.'" The Holy One of Blessing did not redeem only our ancestors, but He even redeemed us with them....[13]

Just as our personal stories are not solely about us but also constitute a major portion of our identity, so too the religious stories constitute a group's identity. The caveat to be added here is that this works one brain at a time, as of course there is not, literally, a group brain to hold a group memory. These stories function in identity formation and maintenance only so long as there are people who repeat them to themselves, their contemporaries, and subsequent generations.

Religion here piggybacks on the way consciousness works. But it does more than constituting self out of our own stories. We draw lessons, construe meaning, based on the group or cultural stories that we take to heart as well as on our personal stories. The renowned Protestant theologian Paul Tillich argued that religions employ myth and symbol "because symbolic expression alone is able to express the ultimate."[14] In premodern times apparently more people took the myths and symbols literally. Non-Christians may wonder at Christian battles, literal as well as intellectual, over whether the bread and wine of the Eucharist are literally or symbolically the body and blood of Christ. That the debate could rage on demonstrates the power of religious story to transcend itself, ceasing to be history—the stories of others—and becoming myth, one's own values expressed in narrative. Myths remain powerful even for those who recognize that many of them are not historical. A myth—for instance, creation in Genesis, or the virgin birth in the Gospels, which many recognize as historically or scientifically impossible—still carries its traditional values. For Tillich, a myth recognized as a myth becomes "a broken myth." Even as a broken myth, however, it may still be an effective vehicle to express its value. Religious believers may or may not believe, for instance, depending on the denomination and the individual, that the world was created in six days, but the repeated refrain that "God saw that it was good" (Genesis 1, vv. 4, 10, 12, 18, 21, 25, 31) expresses a basic attitude about the world. Likewise, the assertion that the human was created in the divine image (Genesis 1:26–27) continues to carry a message about intrinsic human dignity regardless of how creation transpired, over days or eons. Deny that God has a body, and thus a literal image, and every person being "in the image of God" remains powerful as myth, expressing a value and not a fact, a way of regarding people, not of physically describing people. For Tillich, one need not and also must not defend creation, virgin birth, and other myths that defy our scientific knowledge as literal happenings. As myths, they point beyond themselves to deeper truths. We are better off not mistaking penultimate stories and symbols for ultimate reality. "Faith, if it takes its symbols literally, becomes idolatrous! It calls something ultimate which is less than ultimate. Faith, conscious of the symbolic character of its symbols, gives God the honor which is due Him."[15]

One need not necessarily join religious modernists in refusing to take miracle stories literally to recognize that religions use such stories to shape individual and group identity. Looking for criteria for giving one's loyalty to a given tradition beyond providing meaning and motivating ethical behavior (which, as Hick argues, they all do for their adherents), *we may each consider which religious stories, especially as interpreted by each group's clergy and other teachers, seem compelling to us.* That is, does the story of my life indeed seem to echo—as each faith tradition would claim it should—the religious myth or myths about which the faithful cluster? This is, admittedly, a subjective criterion, but of course what those in search of a religion are specifically looking for is a higher paradigm that works for, that helps and satisfies, them. Subjectivity in such a quest is inescapable. It is the goal. A higher paradigm must somehow speak to each of us, must make intuitive sense given our backgrounds, experiences, and personalities.

BEING PASSIONATE ABOUT OUR FAITH

Much though we are apt to find the religious tradition into which we were born compelling, we obviously did not select it after sifting through all the options. By the nature of the "higher paradigms" for which I have been arguing, they are not easily dislodged from the mind. They are, after all, dual mental screens. First, from the inside looking out, they help us construe the world meaningfully through the concepts and categories, the stories and ideals, of the tradition. Second, as other ways of construing meaning try to get in, the screens greatly limit but may not wholly prevent rival paradigms from influencing us. Once indoctrinated (not a negative term—simply the teaching of a doctrine we think will help someone live life), people need an uncommonly powerful reason to change faiths.

But that happens. A woman, we shall call her Suzie, made an appointment to talk to me about converting to Judaism. She explained that in her growing up she often questioned what her pastor or Sunday school instructor was teaching. When she refused to accept what she considered simplistic answers, Suzie recalls being shushed, told she was too young to understand, and sometimes belittled and embarrassed. Perhaps, I suggested, she should try a more liberal variety of Christianity. No, she insisted, she had "gone church-shopping" and long ago decided she simply could not accept certain aspects of Christian doctrine. What she had read about Judaism and discussed with Jewish friends seemed, based on her experience, more appealing.

A man we shall call George, on the other hand, grew up in my synagogue in a reasonably observant home, attended our religious school, and celebrated his bar mitzvah and confirmation with me. When, in his late twenties, his parents asked him to come in and talk to me, I had not seen him in a decade or more. His parents were upset that George was converting to the religion of his Christian fiancée. He explained, a bit apologetically in order not to hurt my feelings, that he had always been bored in our worship services. It was all too cerebral for him. In college he had learned a bit about Eastern religions and meditation. Now, with his fiancée, he found himself inspired, he said, by worship services where it seemed they were actually talking personally with God and learning what the Bible required of them. "Not just 'here is what our texts teach about war or abortion,'" he said, "but 'Thus saith the Lord!'" And they sang in her church so much better than we did in the synagogue! And in a language he understood! There was a greater sense of spirituality, he insisted. George was no more going to be kept in the Jewish fold by my suggesting that there are Jewish forms of meditation, or congregations with more group singing and less choral music, than Suzie was going to be kept in her inherited tradition by reading Tillich. In each case a mix of intellectual and emotional baggage together with new discovery meant to them, in effect, "I was born in that faith, but this one fits my needs better." Once their curiosity had been peaked and enthusiasm for their experiences and learning in the new tradition developed, they began to take the new (to them) religion's stories, rituals, and symbols seriously. Difficult though it may be for Suzie's and George's clergy, who must also comfort bewildered parents, in the free marketplace of ideas in which we live we need to say, "God bless you; go where your heart and mind lead you."

Multiple religions offer different paths to God and different ways of making sense of and thus coping with life. They have different characters and strengths. To live with a religion, as with a spouse, for a lifetime, we need the one that has the greatest appeal to us, both intellectually and emotionally. That is the one we should embrace, not with cool neutrality, but passionately.

But wait! one might object: are we not supposed to be accepting the values a religion teaches, rather than adopting the religion that comes closest to *our* values? Both! Religious values may well have become ours because we were taught them, and thus many people rarely if ever question what they were taught when young. Yet most of us, suffering, observing pain or injustice, or simply maturing and learning to think critically, have found ourselves questioning an inherited belief. We may have sought answers and may or may not have found convincing ones. So a relative few convert to another religion. Ideally all the time, and especially to work *in extremis*, our faith and religious response need to be conditioned, well engrained. A gut-level response, emotional as well as cerebral, will always be more powerful than either emotion or ideas alone. As we saw when considering free will, we most often respond out of our conditioning, yet we are also self-conditioners who seek out further options or generate them when dissatisfied with our first thought.

This, I am convinced, is why there are multiple options within each major religious tradition, as well as why there are multiple traditions. Some crave more freedom and some less. Some appreciate more ritual and some less, more rationalism or more mysticism, greater demands put on them by the community or fewer. The religious communities more authority-based or scripturally literalistic and those more choice-offering and freely interpreting do not end up separate religions altogether. From Orthodox to Reform Judaism, for instance, or Christian denominations from Roman Catholic and Southern Baptist to Presbyterian and United Church of Christ, I would go so far as to suggest that this is because they share the same foundational stories especially and often many of the same rituals. Within each family of religions, we filter the world through the same stories and similar rituals, however differently we may interpret them. Again, in the beginning is the story. Theology is elaborated later. People and cultures are so varied that there can be no one-size-fits-all religion. An efficacious religion is one we can passionately affirm.

We saw above that abstract religious ideas (of salvation, for instance) differ from religious tradition to religious tradition. So, it follows naturally, does the main thrust of religious myths differ. If the most central Christian myth is that of the birth, death, and resurrection of an incarnate God, with its implications for overcoming sin and death through individual faith, and the most central Jewish myth is of the exodus of a people from slavery and degradation to freedom and dignity and on to Sinai with its implications about achieving holiness and progress through corporate action, and the most central Buddhist myth is of a sensitive young man who gives up his royal estate when he discovers the reality of suffering, learning ultimately that suffering is the by-product of desire, and going on to both model and teach that enlightenment can be achieved and desire overcome through meditation and the proper understanding of self and world, we have, to put it mildly, different ways of construing reality!

Such sweeping generalizations obscure points of contact, among them that all three—and other religions, too—are concerned with both individual and communal ethics, that none ignores individual spiritual needs and fear of death, and that each must help individuals deal with suffering. Minority views, moreover, may be found within each.[16] Still, the points of contact should not obscure the differences. Even as no one can *prove* one right and the others wrong, sensitive and serious thinkers will find one or another more *compelling*. Just because we are not God and cannot prove one is superior does not mean, as people, we should not aspire to find the one that makes the most sense to us, a judgment about which there can be no total objectivity, but which will, because of its subjective appeal, help us get more out of the tradition we affirm.

In sum, calling religions "different paths to one God" should not obscure the fact that they are different ways of construing the meaning and purpose of life, history, and all existence. If divine voices came forth from heaven—not just in the stories, but also in each of our personal experience—perhaps we would know with absolute certainty what to believe and what to do. Lacking that, the voices in our sacred traditions may well speak to us. Like Dr. Gazzaniga's split-brain patients,[17] we *will* make sense of our experience. Guided by a religious tradition or some other frame of reference, we will assign meaning. False belief is an option. Nonbelief is not.

At the end of this study, as at the end of each day, I am grateful that the particular combination of matter and energy that is me lives on, perhaps to appreciate more tomorrow. We humans are amazingly blessed. Our brains evolved over the ages, capable of reflecting not only on their environments near and far but also on their own thoughts. Human brains are apprehenders and discoverers of order, assigners of meaning, able to glimpse not only what is but also the significance of what was, what is, and even of what might follow: the patterns, the connections, the purpose, the *meaning*. From each brain's perspective, reality is there to be understood, that we may prosper. So we aspire further, many of us, to catch a glimpse "within" or "behind" reality, a glimpse of God. We are shaped by, but also struggle to shape, reality. We dare to hope we may each contribute something of ourselves to the process: our children, and also ideas and cultural artifacts of all sorts—our brainchildren. Thus the blessing at Jewish weddings, as we note new links being added to the chain of generations and of tradition: *Baruch atah Adonai, Eloheinu Melech ha'olam, shehakol bara lichvodo.* "Blessed are You, Lord our God, Sovereign—Order and Orderer—of Being, who has created everything for Your glory."[18] Amen.

Notes

Introduction

1. *Webster's New World Dictionary of the American Language, College Edition*, ed. Joseph H. Friend and David B. Guralnik (Cleveland and New York: World Publishing, 1957), p. 707.

2. *Webster's New World College Dictionary*, 4th ed., ed. Michael Agnes (Cleveland: Wiley Publishing, 2008), p. 695.

3. ADD is now known as ADHD, H for a common hyperactivity symptom, which, happily, was not pronounced in our son.

1. Our Believing Brains: On Not Being Overwhelmed

1. David J. Linden, *The Accidental Mind: How Brain Evolution Has Given Us Love, Memory, Dreams and God* (Cambridge, MA: Harvard University Press, 2007), p. 28.

2. V. S. Ramachandran, *A Brief Tour of Human Consciousness: From Impostor Poodles to Purple Numbers* (New York: Pi Press, 2004), pp. 7–10.

3. V. S. Ramachandran and Sandra Blakeslee, *Phantoms in the Brain: Probing the Mysteries of the Human Mind* (New York: HarperCollins, 1998), p. 159.

4. Ramachandran, *Brief Tour of Human Consciousness*, p. 8.

5. Oliver Sacks, *The Man Who Mistook His Wife for a Hat and Other Clinical Tales* (New York: Summit Books, 1985), pp. 7–21.

6. Michael S. Gazzaniga, *The Ethical Brain* (New York: Dana Press, 2005), p. 135.

7. Ibid., pp. 136–37.

8. Ibid., pp. 148–49.

2. Taking God Personally

1. See, e.g., the opening chapters of Abraham Joshua Heschel, *God in Search of Man: A Philosophy of Judaism* (Philadelphia: Jewish Publication Society, 1956).

2. See Rudolf Otto, *The Idea of the Holy* (London: Oxford University Press, 1923).

3. See, e.g., Matthew Alper, *The "God" Part of the Brain: A Scientific Interpretation of Human Spirituality and God* (Naperville, IL: Sourcebooks, 2006).

4. Thomas S. Kuhn, *The Structure of Scientific Revolutions* (Chicago: University of Chicago Press, 1962).

5. Although this passage, of course, is biblical—Deuteronomy 6:4—I employ it here to echo its role as the basic affirmation of faith in Jewish prayer.

6. See "Belief in Belief," chap. 8 of Daniel C. Dennett, *Breaking the Spell: Religion as a Natural Phenomenon* (New York: Viking, 2006).

7. Richard Dawkins, *The God Delusion* (Boston: Houghton Mifflin, 2006), p. 19.

8. Dennett, *Breaking the Spell*, p. xiii.

9. Among other passages, Psalm 19:15, Deuteronomy 32:11, and Jeremiah 2:13.

10. Michael S. Gazzaniga, *Human: The Science Behind What Makes Us Unique* (New York: HarperCollins, 2008), p. 49.

11. Ibid., p. 49.

12. Ibid., p. 51.

13. Daniel C. Dennett, *Kinds of Minds: Toward an Understanding of Consciousness* (New York: Basic Books, 1996), p. 27.

14. For a basic statement see chap. 2, "Intentionality: The Intentional Systems Approach," in Dennett, *Kinds of Minds*. Dennett has also written an entire book on the subject, *The Intentional Stance* (Cambridge, MA: MIT Press, 1987).

3. Mystical and Spiritual, Neurological and Theological

1. Author's translation.

2. Joseph Dan, *The Heart and the Fountain: An Anthology of Jewish Mystical Experiences* (Oxford: Oxford University Press, 2002), p. 3.

3. *Union Prayer Book*, newly rev. ed. (Cincinnati: Central Conference of American Rabbis, 1940), p. 7.

4. *Leviticus Rabbah* 24:5.

5. Christine Wicker, "How Spiritual Are We?" *Parade*, October 4, 2009, pp. 4–5.

6. In Christianity this desire may be satisfied at least in part by the Eucharist. Those who take that ritual literally as transubstantiation believe themselves to be tasting God. Christians who understand "body and blood" as symbolic, not literal, however, are in the same position as Jews, Muslims, and others who think of God as utterly nonmaterial. The long history of Catholic mysticism demonstrates that even the Eucharist has not always been sufficient to satisfy those who yearn for experience of the Divine somehow deeper than taste.

7. William James, *The Varieties of Religious Experience* (New York: Barnes & Noble Classics, text based on 1902 ed., 2004), pp. 328–71.

8. See Moshe Idel, *Kabbalah: New Perspectives* (New Haven, CT: Yale University Press, 1988), pp. ix–xix.

9. Sigmund Freud, in *Civilization and Its Discontents*, p. 11, quotes Romain Rolland using this description of the mystic state. Cited in Matthew Alper, *The "God" Part of the Brain* (Naperville, IL: Sourcebooks, 2006), pp. 131–32.

10. Dan, *The Heart and the Fountain*, pp. 1–7.

11. Andrew Newberg, Eugene d'Aquili, and Vince Raus, *Why God Won't Go Away: Brain Science and the Biology of Belief* (New York: Ballantine Books, 2001), p. 2.

12. Ibid., pp. 4–5.

13. Andrew Newberg and Mark Robert Waldman, *Why We Believe What We Believe: Uncovering Our Biological Need for Meaning, Spirituality, and Truth* (New York: Free Press, 2006), p. 173.

14. For a more detailed anatomical description, see ibid., pp. 174–80.

15. Jill Bolte Taylor, *My Stroke of Insight: A Brain Scientist's Personal Journey* (New York: Viking, 2008).

16. Ibid., pp. 29–30.

17. Ibid., p. 68.

18. Ibid., pp. 67–68.

19. Ibid., p. 69.

20. Antonio Damasio, *The Feeling of What Happens: Body and Emotion in the Making of Consciousness* (San Diego, CA: Harcourt, 1999), especially chap. 2, "Emotion and Feeling," pp. 35–81, and chap. 9, "Feeling Feelings," pp. 279–95.

21. J. L. Saver and J. Rabin, "The Neural Substrates of Religious Experience," *Journal of Neuropsychology and Clinical Neuroscience*, 9, no. 3 (1997): 498–510 (cited in Newberg and Waldman, *Why We Believe*, chap. 7, n. 25).

22. Nina P. Azari et. al., "Neural Correlates of Religious Experience," *European Journal of Neuroscience*, 13, no. 8 (2001): 1649–52.

23. "Creating Emotional Realities," in Newberg and Waldman, *Why We Believe*, pp. 180–83.

24. Malcolm Jeeves and Warren S. Brown, *Neuroscience, Psychology, and Religion: Illusions, Delusions, and Realities about Human Nature* (West Conshohocken, PA: Templeton Press, 2009), pp. 95–97.

25. Damasio, *Feeling of What Happens*, p. 51.

26. Martin Buber, *Tales of the Hasidim: Early Masters* (New York: Schocken Books, 1947), pp. 246–47.

27. Lawrence A. Hoffman, *Israel: A Spiritual Travel Guide* (Woodstock, VT: Jewish Lights, 2005), pp. 158–59.

4. The Soul Which Thou Hast Given unto Me?

1. *Union Prayer Book,* newly rev. ed. (Cincinnati: Central Conference of American Rabbis, 1940), p. 101.

2. Jon Levenson, *Resurrection and the Restoration of Israel: The Ultimate Victory of the God of Life* (New Haven, CT: Yale University Press, 2006), p. 166.

3. Funeral liturgies, in *Rabbi's Manual* (New York: Central Conference of American Rabbis, 1961), p. 74; *Rabbi's Manual*, ed. David Polish (New York: Central Conference of American Rabbis, 1988), p. 159.

4. *Authorized Daily Prayer Book*, rev. ed., ed. Joseph Hertz (New York: Bloch, 1963), pp. 18–19.

5. For a recent example, see *Union Prayer Book*, p. 44, but the wording goes back to nineteenth-century Reform rabbi David Einhorn; see Neil Gillman, *The Death of Death: Resurrection and Immortality in Jewish Thought* (Woodstock, VT: Jewish Lights, 1997), p. 201.

6. *Gates of Prayer*, ed. Chaim Stern (New York: Central Conference of American Rabbis, 1975), p. 38.

7. *Mishkan T'filah*, ed. Elyse D. Frishman (New York: Central Conference of American Rabbis, 2007), p. 78.

8. *Sabbath and Festival Prayer Book*, Silverman (1946) and *Siddur Sim Shalom* (1985), quoted in Gillman, *Death of Death*, pp. 205–7.

9. *Mishkan T'filah*, p. 593.

10. *Mishkan T'filah*, p. 595.

11. See Revised Standard Version (1901) and *Tanakh* (Philadelphia: Jewish Publication Society, 1985), *inter alia*.

12. *Encyclopaedia Judaica*, 2nd ed. (Detroit, MI: Thomson Gale/ Macmillan Reference, 2007), v. 19, p. 33; Jack Bemporad, "Soul, Jewish Concepts," *Encyclopedia of Religion*, vol. 13, pp. 450–55; Kaufmann Kohler, *Jewish Theology, Systematically and Historically Considered* (New York: Macmillan, 1918), p. 212.

13. Jan Bremmer, "Soul, Greek and Hellenistic Concepts," *Encyclopedia of Religion*, vol. 13, p. 434.

14. Frank Thilly and Ledger Wood, *A History of Philosophy*, 3rd ed. (New York: Holt, Rinehart and Winston, 1956), p. 85.

15. Ibid., p. 86.

16. Eric R. Kandel, *In Search of Memory: The Emergence of a New Science of Mind* (New York: Norton, 2006), p. 215.

17. This translation from *Mishkan T'filah*, p. 194; or see *Authorized Daily Prayer Book*, ed. Hertz, p. 10.

18. Daniel C. Dennett, *Freedom Evolves* (New York: Viking, 2003), p. 249.

19. Denis Dutton, *The Art Instinct: Beauty, Pleasure and Human Evolution* (New York: Bloomsbury Press, 2009), pp. 141–42.

20. V. S. Ramachandran, *A Brief Tour of Human Consciousness: From Impostor Poodles to Purple Numbers* (New York: Pi Press, 2004), pp. 21–22.

21. V. S. Ramachandran and Sandra Blakeslee, *Phantoms in the Brain: Probing the Mysteries of the Human Mind* (New York: Quill, 1999), pp. 199–201.

22. For this and much more, see Daniel J. Levitan, *This Is Your Brain on Music: The Science of a Human Obsession* (New York: Dutton, 2006).

23. Dutton, *The Art Instinct*, p. 156. The entire chapter, "Art and Human Self-Domestication," develops the theme of sexual selection and the arts. Where music in particular is concerned, also see Levitan, *This Is Your Brain on Music*, pp. 245–52.

24. *Gates of Prayer*, p. 624.

25. "Holy Sonnets 10," often known as "Death Be Not Proud."

26. See Gillman, *The Death of Death*, especially the final chapter, "What Do I Believe?" and its conclusion on pp. 273–74.

27. Richard Elliot Friedman, *Who Wrote the Bible?* (New York: Summit, 1987), p. 272, n. 14.

28. Jon D. Levenson, *Resurrection and the Restoration of Israel: The Ultimate Victory of the God of Life* (New Haven, CT: Yale University Press, 2006), chaps. 3 and 4; quotation on p. 78.

5. Free Will and Free Won't: Programming Your Brain

1. The first time I heard this story it was told to me as a Jewish tale, with the sage a rabbi. An Internet search turned up several versions of the tale, with no clear source, some identified as Asian, with a male sage, and some as Native American, with a female sage.

2. Philo, "On the Confusion of Tongues," XXXV, *The Works of Philo Judaeus*, trans. Charles Duke Yonge (London: H. G. Bohn, 1854–90; available on several websites, including www. earlyjewishwritings.com).

3. For reinterpretations of Augustine's idea of original sin, consider Reinhold Niebuhr and Paul Tillich. Niebuhr wrote that the idea of *inherited* original sin is an error of Augustinian literalism; the Garden of Eden story is not the issue. Original sin is part of our nature, a "dialectical truth," for "self-love and self-centeredness is inevitable, but not in such a way as to fit into the category of natural necessity" (*The Nature and Destiny of Man: A Christian Interpretation*, vol. 1 [New York: Charles Scribner's Sons, 1941], chap. 9, quotation on p. 263). Tillich speaks of "reinterpret[ing] the doctrine of original sin by showing man's existential self-estrangement" and even flirts with "the definite removal from the theological vocabulary of terms like 'original sin' or 'hereditary sin' and their replacement by a description of the interpenetration of the moral and the tragic elements in the human situation" (*Systematic Theology*, vol. 2 [Chicago: University of Chicago Press, 1957], pp. 38–39).

4. Traditional *Vidui*, this translation from *Gates of Repentance* (New York: Central Conference of America Rabbis, 1978), p. 269.

5. See, for instance, Talmud, *Sukkah* 52b, where God is said to repent having created four things: exile, the Chaldeans, the Ishmaelites, and the *yetzer hara*. Similarly, *Genesis Rabbah* 27:4 says, "It was a regrettable error on My part to have created a *yetzer hara* within him, for had I not created a *yetzer hara*, he would not have rebelled against Me."

6. Soncino Press translation, 1935. Other versions of the parable may be found in midrash; see *Leviticus Rabbah* 4:5.

7. Theories of consciousness and decision making are fascinating, but would be a distraction here. See Daniel Dennett on multiple draft theory in *Consciousness Explained* (Boston: Little Brown, 1991), chap. 5; and Antonio Damasio on somatic marker hypothesis in *Descartes' Error* (New York: Grosset/Putnam, 1994), chap. 8.

8. Daniel C. Dennett, *Freedom Evolves* (New York: Viking, 2003), pp. 227–30; V. S. Ramachandran, *A Brief Tour of Human Consciousness* (New York: Pi Press, 2004), pp. 86–87.

9. Cited in Dennett, *Freedom Evolves*, p. 231.

10. Dennett, *Freedom Evolves*, p. 238.

11. Daniel M. Wegner, *The Illusion of Conscious Will* (Cambridge, MA: MIT Press, 2002).

12. Ibid., p. 325.

13. Ibid., p. 9–10.

14. Ibid., pp. 63–64.

15. Ibid., p. 328.

16. Jill Bolte Taylor, *My Stroke of Insight: A Brain Scientist's Personal Journey* (New York: Viking, 2006), p. 146.

17. In his now classic psychology experiment, Dr. Harry Harlow of the University of Wisconsin built surrogate mothers for rhesus monkeys out of cloth and out of wire mesh. Even when the mesh mother provided milk and the cloth mother did not, the infant monkeys preferred the soft cloth one, and those restricted to contact with the wire one failed to thrive. Variations on the experiment have yielded similar results. See Tiffany Field, *Touch* (Cambridge, MA: MIT Press, 2001), pp. 35–36; or H. F. Harlow and R. R. Zimmerman, "The Development of Affectionate Responses in Infant Monkeys," *Proceedings of the American Philosophical Society* 102 (1958): 501–509. Field, director of touch research at the University of Miami School of Medicine, discusses the importance of touch for people at various stages of our development, including in sexuality. See especially, "Touch in Adulthood," *Touch*, pp. 53–57.

18. See note 4.

6. Morality: The Hop of Faith

1. Rabbi Richard Levy, meditation on theme of *Ahavat Olam* prayer, *Gates of Prayer* (New York: Central Conference of American Rabbis, 1975), pp. 249–50.

2. In Judaism, for instance, the Torah distinguishes between what we would call accidental or negligent death, and thus manslaughter, and intentional murder (Numbers 35:9–28). Where the crime is clearly murder, monetary compensation may not be substituted for capital punishment (Numbers 35:31), which implies that monetary payment may be substituted for lesser injuries or perhaps even other capital crimes, of which the Hebrew Bible has many. Numbers 35:33 likewise states that the blood pollution of the land caused by a murder can be expiated by nothing less than the death of the murderer. Struggling against this absolute insistence on death as the only fitting punishment for murder, we find great controversy among first-century sages in *Mishnah Makkot* 1:10. Following rules of evidence that appear intended to make it very difficult for a court to render a capital sentence, the Mishnah states that a Sanhedrin that puts someone to death every seven years—or Rabbi Eliezer ben Azariah says, seventy years!—is considered "destructive," overly harsh. Rabbis Tarphon and Akiva say that "had we been in the Sanhedrin none would ever have been put to death," and Rabbi Simeon ben Gamliel responds that those sages would thus have "multiplied the shedders of blood in Israel."

3. For example, can anything but a defensive war—if that—be morally justified? In the Hebrew Bible aggressive war is actually commanded. God instructs the Hebrews to fight Amalek and other enemies and to conquer Canaan, though Rabbinic thought refuses to make precedents of those biblical struggles, declaring, for instance, that the seven enemy nations of the Bible have assimilated and may no longer be fought. Only defensive war remains as *milchemet mitzvah* (divinely commanded war), though some consider a preemptive strike defensive. In Christian thought a strong pacifist tradition that holds all wars to be immoral coexists with an extensive tradition that attempts to prescribe criteria and thus guidelines for the "just war," in which, for instance, war should only be waged

as a last result, for a just cause, with the anticipated benefits proportionate to the likely suffering and destruction, and so on. Even within our moral traditions, much less between them, we do not agree.

4. For a thoughtful and entertainingly written survey of such philosophies, with provocative contemporary examples of how they apply to events and differ with one another, see Michael Sandel, *Justice: What's the Right Thing to Do?* (New York: Farrar, Straus and Giroux, 2009).

5. The roots of biblical criticism, of course, go back far beyond the nineteenth century when it triumphed in the academies of higher learning. It would take us too far afield to go back to Spinoza's *Tractatus Theologico-Politicus* in seventeenth-century Holland, Abraham ibn Ezra's biblical commentaries in twelfth-century Spain, or other early sources.

6. For scripture as occasion for sensing God's presence in life, see the discussion of spirituality in chap. 3.

7. Philippa Foot, "The Problem of Abortion and the Doctrine of the Double Effect," *Oxford Review* 5, nos. 8–9 (1967), reprinted in Foot, *Virtues and Vices and Other Essays in Moral Philosophy* (Berkeley: University of California Press, 1978), pp. 19 and 23.

8. Joshua Greene et al., "An fMRI Investigation of Emotional Engagement in Moral Judgment," *Science* 293 (2001): 2105–8.

9. Marc D. Hauser, *Moral Minds: How Nature Designed Our Universal Sense of Right and Wrong* (New York: HarperCollins, 2006), p. 33.

10. Greene et al., "An fMRI Investigation."

11. Hauser, *Moral Minds*, p. 112.

12. Kwame Anthony Appiah, *Experiments in Ethics* (Cambridge, MA: Harvard University Press, 2008), p. 89.

13. Philosopher Patricia Churchland questions whether such hypothetical examples are sufficiently like real-world cases for people's responses to produce meaningful data. I remain unconvinced by her argument and would counter that the removal of secondary details, so that it is clear what respondents are reacting to, is precisely what makes the results meaningful. Churchland acknowledges that the research may

show that innate morality is consistent with the data but argues that that remains short of conclusively proving the existence of innate morality, because she can think of other possible explanations. See her *Braintrust: What Neuroscience Tells Us about Morality* (Princeton, NJ: Princeton University Press, 2011), pp. 104–11.

14. Jonathan Haidt, *The Happiness Hypothesis: Finding Modern Truth in Ancient Wisdom* (New York: Basic Books, 2006), pp. 20–21.

15. Hauser, "Justice for All," in *Moral Minds*, pp. 59–110. (I should note that though Hauser refers to multiple sources, including Dr. John Mikhail, Hauser has recently been accused, but denies, taking more ideas from Mikhail than he acknowledges. For our purposes here it suffices to observe that the validity of the *Moral Minds* arguments is not at issue in the case, but only their origin. [viz. p. 32 of Charles Gross, "Disgrace," in *The Nation*, January 9/16, 2012, pp. 25–32.])

16. Hauser, *Moral Minds*, p. 79.

17. Ibid., p. 165.

18. The psychopath, lacking the moral instincts others have, is the relatively rare exception; ibid., pp. 232–41.

19. "The word psychopath describes a person who lacks a conscience, the ability or sensitivity to shape a moral issue and to understand the significance of such issues, and to experience empathy" (Laurence Tancredi, *Hardwired Behavior: What Neuroscience Reveals about Morality* [New York: Cambridge University Press, 2005], p. 46). Tancredi, a New York University clinical professor of psychiatry, explains that research has yet to establish how much psychopathology is "nature" and how much "nurture," but it is both—a genetic potential may be present without environment activating it. Thus when two siblings grow up in the same abusive household, one may turn out to be a psychopath while the other does not (ibid., pp. 46–68).

20. Frans de Waal, *Primates and Philosophers* (Princeton, NJ: Princeton University Press, 2006), p. 14.

21. Frans de Waal, *Our Inner Ape* (New York: Riverhead Books, 2005), pp. 44–47 and 79–80.

22. De Waal, *Primates*, p. 26.

23. De Waal, *Our Inner Ape*, p. 6.

24. Frans de Waal, *The Age of Empathy* (New York: Harmony Books, 2009), p. 10–11.

25. De Waal, *Our Inner Ape*, p. 30; *Primates*, pp. 33–36.

26. De Waal, *Primates*, p. 15.

27. De Waal's basic case in remarkably concise form may be found in *Primates*, with many more, often entertaining, anecdotes in *Our Inner Ape* and *The Age of Empathy*.

28. De Waal, *The Age of Empathy*, p. 172; Hauser, *Moral Minds*, pp. 383–85.

29. De Waal, *Primates*, p. 18.

30. De Waal, *The Age of Empathy*, p. 43.

31. De Waal, *Primates*, pp. 39–40; *The Age of Empathy*, pp. 208–9.

32. Daniel C. Dennett, *Darwin's Dangerous Idea: Evolution and the Meaning of Life* (New York: Simon and Schuster, 1995), p. 127.

33. Dennett, "Forced Moves in the Game of Design," in *Darwin's Dangerous Idea*, pp. 128–35.

34. Appiah, *Experiments in Ethics*, p. 98.

7. Life Is with People: Organized Religion

1. Steven R. Quartz and Terrence J. Sejnowski, *Liars, Lovers, and Heroes: What the New Brain Science Reveals about How We Become Who We Are* (New York: HarperCollins, 2002), p. 265.

2. Ibid., pp. 265–66.

3. Ibid., p. 256.

4. Robert D. Putnam, *Bowling Alone: The Collapse and Revival of American Community* (New York: Simon & Schuster, 2000).

5. Ibid., p. 54.

6. Ibid., pp. 60–61.

7. Ibid., p. 69.

8. Ibid., p. 74.

9. Ibid., p. 19.

10. Azriel Eisenberg, *The Synagogue through the Ages* (New York: Bloch, 1974), pp. 41–42.

11. Michael Avi-Yonah, "Synagogue, Architecture," in *Encyclopaedia Judaica*, 2nd ed. (Detroit, MI: Thomson Gale/ Macmillan Reference, 2007), vol. 19, p. 365.

12. Martin Buber, *I and Thou*, trans. Ronald Gregor Smith (New York: Charles Scribner's Sons, 1958), p. 11.

13. I recall my graduate school philosophy professor, Dr. Alvin Reines, telling this story about George Santayana. I have never found it in print, so it may be apocryphal, but it makes the point.

14. For a thoughtful summary of how such factors influence worship services, see Lawrence Hoffman, *The Art of Public Prayer: Not for Clergy Only*, 2nd ed. (Woodstock, VT: SkyLight Paths, 1999). A series of Hoffman studies on how worship works has greatly influenced my thinking over the years.

15. Clifford Geertz, *The Interpretation of Cultures* (New York: Basic Books, 1973), pp. 87–141.

16. Adam B. Seligman, Robert P. Weller, Michael J. Puett, and Bennett Simon, *Ritual and Its Consequences: An Essay on the Limits of Sincerity* (New York: Oxford University Press, 2008).

17. Andrew Newberg, Eugene d'Aquili, and Vince Rause, *Why God Won't Go Away* (New York: Random House, 2001), p. 89. Chap. 5, "Ritual: The Physical Manifestation of Meaning," is a worthwhile summary of the interaction of ideas, actions, and neurology in public worship.

18. Mordecai M. Kaplan, *Questions Jews Ask: Reconstructionist Answers* (New York: Reconstructionist Press, 1956), pp. 102–3.

19. The title of this chapter echoes the title of Mark Zborowski's and Elizabeth Herzog's classic anthropological study, *Life Is with People: The Culture of the Shtetl* (Madison, CT: International Universities Press, 1951).

8. Why My Religion? What of Yours?

1. Told in Gershom Scholem, *Major Trends in Jewish Mysticism* (Jerusalem: Schocken, 1941), pp. 349–50.

2. Thomas Lewis, Fari Amini, and Richard Lennon, *A General Theory of Love* (New York: Vintage/Random House, 2001), p. 68.

3. John Hick, *The New Frontier of Religion and Science: Religious Experience, Neuroscience and the Transcendent* (New York: Palgrave Macmillan, 2010).

4. Ibid., p. 146.

5. Ibid., p. 148.

6. Ibid., p. 150.

7. Ibid., p. 149.

8. Hick notes that he is using this Kantian insight in his own way; "Kant himself was discussing sense perception and did not apply the distinction to religion" (ibid., pp. 138, 163–64; quotation on p. 163).

9. In Judaism we also have a mythic sense of being one people, an ethnic component as well as a more purely religious one.

10. From *Aleinu* rubric of the daily Jewish service.

11. This is the title of Hick's fourth chapter.

12. Daniel Dennett, *Consciousness Explained* (Boston: Little, Brown, 1991), p. 418; emphasis in original.

13. Translation by Joel Hoffman, in *My People's Passover Haggadah*, ed. Lawrence A. Hoffman and David Arnow (Woodstock, VT: Jewish Lights, 2008), vol. 2, p. 76., emphasis added.

14. Paul Tillich, *Dynamics of Faith* (New York: Harper and Row, 1957), p. 41.

15. Ibid., pp. 51–52.

16. Years ago Rabbi Lawrence Kushner, in response to a colleague who thought a Hasidic teaching sounded very Buddhist, told a class in which I was participating, "There are only a few big religious ideas, and every tradition gets around to all of them eventually."

17. See chap. 1, n. 8.

18. Second blessing of the *Sheva Berachot*, the traditional seven wedding blessings of Jewish liturgy, with my interpretive translation.

Suggested Readings and Selected Bibliography

Readers who want to delve further into cognitive studies topics explored in this book should find the selected bibliography useful. It is a listing of books that I have found helpful as I have explored cognitive science. Details on other works cited in the text, of course, are available in the notes.

From time to time people whose curiosity is peaked by one or more of the areas in this book ask me what they should read, looking not for a lengthy list of possibilities, but simply for a provocative introduction accessible to the nonspecialist. In this burgeoning field not even a specialist could have read every available book, and I have not, either. I am apt to recommend what I have found readable as well as helpful—inevitably a subjective and even idiosyncratic selection. So I apologize to authors of meritorious books omitted from this list and urge readers to think of these books as simply a place to begin further exploration.

Suggested Readings

1. Our Believing Brains: On Not Being Overwhelmed

Damasio, Antonio. *The Feeling of What Happens: Body and Emotion in the Making of Consciousness*. San Diego: Harcourt, 1999. Dense but valuable reading, this time by a physician-scientist with an appreciation of philosophy.

Dennett, Daniel C. *Consciousness Explained*. Boston: Little, Brown, 1991. Heavy reading by a philosopher with an appreciation of science, but worth the effort.

Eagleman, David. *Incognito: The Secret Lives of the Brain*. New York: Pantheon, 2011. Entertainingly written as well as enlightening; a great place to begin.

Ramachandran, V. S. *A Brief Tour of Human Consciousness: From Impostor Poodles to Purple Numbers*. New York: Pi Press, 2004. Based on a series of public lectures; another wonderful place to begin.

2. Taking God Personally

Gazzaniga, Michael S. *Human: The Science Behind What Makes Us Unique*. New York: HarperCollins, 2008. Intentionality and theory of mind—and many other areas—are nicely explained.

3. Mystical and Spiritual, Neurological and Theological

Damasio, Antonio. *Looking for Spinoza: Joy, Sorrow, and the Feeling Brain*. Orlando, FL: Harcourt, 2003. Any of the Damasio books in the bibliography will do for the emotional element in thinking; this one even gets around to some speculation on spirituality.

Newberg, Andrew, Eugene d'Aquili, and Vince Raus. *Why God Won't Go Away: Brain Science and the Biology of Belief*. New York: Ballantine Books, 2001. Includes taking pictures of meditating brains—and more broadly introducing "neurotheology."

Taylor, Jill Bolte. *My Stroke of Insight: A Brain Scientist's Personal Journey*. New York: Viking, 2008. In addition to providing a first-person echo of Newberg, d'Aquili, and Raus, this should be mandatory reading for anyone working with stroke patients.

4. The Soul Which Thou Hast Given unto Me?

Dutton, Denis. *The Art Instinct: Beauty, Pleasure and Human Evolution*. New York: Bloomsbury Press, 2009. For evolution and art.

Levitan, Daniel J. *This Is Your Brain on Music: The Science of a Human Obsession*. New York: Dutton, 2006. For evolution and music.

Ramachandran, V. S. *The Tell-Tale Brain: A Neuroscientist's Quest for What Makes Us Human*. New York: Norton, 2011. Chapters 7 and 8 explore the neurology of art appreciation, and other chapters are a nice review of Ramachandran's earlier books.

5. Free Will and Free Won't: Programming Your Brain

Dennett, Daniel C. *Freedom Evolves*. New York: Viking, 2003. A basic introduction to free will and the brain.

Wegner, Daniel M. *The Illusion of Conscious Will*. Cambridge, MA: MIT Press, 2002. What, scientifically, is a "will"? I discovered this book in the course of reading the previous one, and I second Dennett's observation that Wegner's title is "awkward" (*Freedom Evolves*, p. 224), for Wegner does *not* argue that we are not free moral agents.

6. Morality: The Hop of Faith

de Waal, Frans. *Primates and Philosophers*. Princeton, NJ: Princeton University Press, 2006. De Waal's basic insights into the evolution of human morality from animal behavior are nicely summarized in a mere fifty pages (there are also critiques by others). But you may find de Waal's other two books in the bibliography below, enriched with anecdotes from his vast experience, more entertaining.

Hauser, Marc D. *Moral Minds: How Nature Designed Our Universal Sense of Right and Wrong*. New York: HarperCollins, 2006. Long, but only because he has so much to say!

7. Life Is with People: Organized Religion

Quartz, Steven R., and Terrence J. Sejnowski. *Liars, Lovers, and Heroes: What the New Brain Science Reveals about How*

We Become Who We Are. New York: HarperCollins, 2002. The research I employ on brain size preparing us to function optimally in groups of about 150 people is Robin Dunbar's, and Dunbar recently came out with his own book (see selected bibliography). Both he and Quartz and Sejnowski cover many other topics, so for the whole book, as opposed to specifically the section on brain size, I recommend this one more highly.

8. Why My Religion? What of Yours?

Dennett, Daniel C. *Consciousness Explained.* Boston: Little, Brown, 1991. For narrative self as a source of identity, see particularly chap. 13, "The Reality of Selves."

Gazzaniga, Michael S. *Human: The Science Behind What Makes Us Unique.* New York: HarperCollins, 2008. Somewhat lighter in style than Dennett, Gazzaniga also covers narrative self, particularly in chap. 8, "Is Anybody There?"

Selected Bibliography

Alper, Matthew. *The "God" Part of the Brain: A Scientific Interpretation of Human Spirituality and God.* Naperville, IL: Sourcebooks, 2006.

Appiah, Kwame Anthony. *Experiments in Ethics.* Cambridge, MA: Harvard University Press, 2008.

Bok, Sissela. *Exploring Happiness: From Aristotle to Brain Science.* New Haven, CT: Yale University Press, 2010.

Churchland, Patricia. *Braintrust: What Neuroscience Tells Us about Morality.* Princeton, NJ: Princeton University Press, 2011.

Damasio, Antonio. *Descartes' Error: Emotion, Reason, and the Human Brain.* New York: Grosset/Putnam, 1994.

———. *The Feeling of What Happens: Body and Emotion in the Making of Consciousness.* San Diego, CA: Harcourt, 1999.

———. *Looking for Spinoza: Joy, Sorrow, and the Feeling Brain.* Orlando, FL: Harcourt, 2003.

Dennett, Daniel C. *Consciousness Explained.* Boston: Little, Brown, 1991.

———. *Darwin's Dangerous Idea: Evolution and the Meaning of Life.* New York: Simon and Schuster, 1995.

———. *Freedom Evolves*. New York: Viking, 2003.

———. *Kinds of Minds: Toward an Understanding of Consciousness*. New York: Basic Books, 1996.

———. *Sweet Dreams: Philosophical Obstacles to a Science of Consciousness*. Cambridge, MA: MIT Press, 2005.

de Waal, Frans. *The Age of Empathy*. New York: Harmony Books, 2009.

———. *Our Inner Ape*. New York: Riverhead Books, 2005.

———. *Primates and Philosophers*. Princeton, NJ: Princeton University Press, 2006.

Dunbar, Robin. *How Many Friends Does One Person Need? Dunbar's Number and Other Evolutionary Quirks*. Cambridge, MA: Harvard University Press, 2010.

Dutton, Denis. *The Art Instinct: Beauty, Pleasure and Human Evolution*. New York: Bloomsbury Press, 2009.

Eagleman, David. *Incognito: The Secret Lives of the Brain*. New York: Pantheon, 2011.

Gazzaniga, Michael S. *The Ethical Brain*. New York: Dana Press, 2005.

———. *Human: The Science Behind What Makes Us Unique*. New York: HarperCollins, 2008.

Gilbert, Daniel. *Stumbling on Happiness*. New York: Alfred A. Knopf, 2006.

Haidt, Jonathan. *The Happiness Hypothesis: Finding Modern Truth in Ancient Wisdom*. New York: Basic Books, 2006.

Hauser, Marc D. *Moral Minds: How Nature Designed Our Universal Sense of Right and Wrong*. New York: HarperCollins, 2006.

Hick, John. *The New Frontier of Religion and Science: Religious Experience, Neuroscience and the Transcendent*. New York: Palgrave Macmillan, 2010.

Hogue, David A. *Remembering the Future, Imagining the Past: Story, Ritual, and the Human Brain*. Cleveland: Pilgrim Press, 2003.

Jeeves, Malcolm, and Warren S. Brown. *Neuroscience, Psychology, and Religion: Illusions, Delusions, and Realities about Human Nature*. West Conshohocken, PA: Templeton Foundation Press, 2009.

Joyce, Richard. *The Evolution of Morality*. Cambridge, MA: MIT Press, 2007.

Kandel, Eric R. *In Search of Memory: The Emergence of a New Science of Mind*. New York: Norton, 2006.

Levitan, Daniel J. *This Is Your Brain on Music: The Science of a Human Obsession*. New York: Dutton, 2006.

Lewis, Thomas, Fari Amini, and Richard Lennon. *A General Theory of Love*. New York: Vintage/Random House, 2001.

Linden, David J. *The Accidental Mind: How Brain Evolution Has Given Us Love, Memory, Dreams and God*. Cambridge, MA: Harvard University Press, 2008.

Newberg, Andrew, Eugene d'Aquili, and Vince Raus. *Why God Won't Go Away: Brain Science and the Biology of Belief*. New York: Ballantine Books, 2001.

Newberg, Andrew, and Mark Robert Waldman. *Why We Believe What We Believe: Uncovering Our Biological Need for Meaning, Spirituality, and Truth*. New York: Free Press, 2006.

Pinker, Steven. *The Blank Slate: The Modern Denial of Human Nature*. New York: Penguin Books, 2002.

Quartz, Steven R., and Terrence J. Sejnowski. *Liars, Lovers, and Heroes: What the New Brain Science Reveals about How We Become Who We Are*. New York: HarperCollins, 2002.

Ramachandran, V. S. *A Brief Tour of Human Consciousness: From Impostor Poodles to Purple Numbers*. New York: Pi Press, 2004.

———. *The Tell-Tale Brain: A Neuroscientist's Quest for What Makes Us Human*. New York: Norton, 2011.

——— and Sandra Blakeslee. *Phantoms in the Brain: Probing the Mysteries of the Human Mind*. New York: HarperCollins, 1998.

Rose, Steven. *The Future of the Brain: The Promise and Perils of Tomorrow's Neuroscience*. New York: Oxford University Press, 2005.

Sacks, Oliver. *The Man Who Mistook His Wife for a Hat and Other Clinical Tales*. New York: Summit Books, 1985.

Sapolsky, Robert M. *Monkeyluv*. New York: Scribner, 2005.

Tancredi, Laurence. *Hardwired Behavior*. New York: Cambridge University Press, 2005.

Index

189

Theology/Philosophy

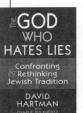

The God Who Hates Lies: Confronting & Rethinking Jewish Tradition
By Dr. David Hartman with Charlie Buckholtz
The world's leading Modern Orthodox Jewish theologian probes the deepest questions at the heart of what it means to be a human being and a Jew.
6 x 9, 208 pp, HC, 978-1-58023-455-9 **$24.99**

Jewish Theology in Our Time: A New Generation Explores the Foundations and Future of Jewish Belief *Edited by Rabbi Elliot J. Cosgrove, PhD; Foreword by Rabbi David J. Wolpe; Preface by Rabbi Carole B. Balin, PhD*
A powerful and challenging examination of what Jews can believe—by a new generation's most dynamic and innovative thinkers.
6 x 9, 240 pp, HC, 978-1-58023-413-9 **$24.99**

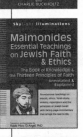

Maimonides—Essential Teachings on Jewish Faith & Ethics: The Book of Knowledge & the Thirteen Principles of Faith—Annotated & Explained
Translation and Annotation by Rabbi Marc D. Angel, PhD
Opens up for us Maimonides's views on the nature of God, providence, prophecy, free will, human nature, repentance and more.
5½ x 8½, 224 pp, Quality PB Original, 978-1-59473-311-6 **$18.99***

The Death of Death: Resurrection and Immortality in Jewish Thought
By Rabbi Neil Gillman, PhD 6 x 9, 336 pp, Quality PB, 978-1-58023-081-0 **$18.95**

Doing Jewish Theology: God, Torah & Israel in Modern Judaism *By Rabbi Neil Gillman, PhD*
6 x 9, 304 pp, Quality PB, 978-1-58023-439-9 **$18.99**

Hasidic Tales: Annotated & Explained *Translation & Annotation by Rabbi Rami Shapiro*
5½ x 8½, 240 pp, Quality PB, 978-1-893361-86-7 **$16.95***

A Heart of Many Rooms: Celebrating the Many Voices within Judaism
By Dr. David Hartman 6 x 9, 352 pp, Quality PB, 978-1-58023-156-5 **$19.95**

The Hebrew Prophets: Selections Annotated & Explained
Translation & Annotation by Rabbi Rami Shapiro; Foreword by Rabbi Zalman M. Schachter-Shalomi
5½ x 8½, 224 pp, Quality PB, 978-1-59473-037-5 **$16.99***

Maimonides, Spinoza and Us: Toward an Intellectually Vibrant Judaism
By Rabbi Marc D. Angel, PhD A challenging look at two great Jewish philosophers and what their thinking means to our understanding of God, truth, revelation and reason. 6 x 9, 224 pp, HC, 978-1-58023-411-5 **$24.99**

A Living Covenant: The Innovative Spirit in Traditional Judaism
By Dr. David Hartman 6 x 9, 368 pp, Quality PB, 978-1-58023-011-7 **$25.00**

Love and Terror in the God Encounter: The Theological Legacy of Rabbi Joseph
B. Soloveitchik *By Dr. David Hartman* 6 x 9, 240 pp, Quality PB, 978-1-58023-176-3 **$19.95**

A Touch of the Sacred: A Theologian's Informal Guide to Jewish Belief
By Dr. Eugene B. Borowitz and Frances W. Schwartz
6 x 9, 256 pp, Quality PB, 978-1-58023-416-0 **$16.99**; HC, 978-1-58023-337-8 **$21.99**

Traces of God: Seeing God in Torah, History and Everyday Life *By Rabbi Neil Gillman, PhD*
6 x 9, 240 pp, Quality PB, 978-1-58023-369-9 **$16.99**

Your Word Is Fire: The Hasidic Masters on Contemplative Prayer
Edited and translated by Rabbi Arthur Green, PhD, and Barry W. Holtz
6 x 9, 160 pp, Quality PB, 978-1-879045-25-5 **$15.95**

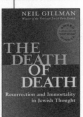

Or phone, fax, mail or e-mail to: **JEWISH LIGHTS Publishing**
Sunset Farm Offices, Route 4 • P.O. Box 237 • Woodstock, Vermont 05091
Tel: (802) 457-4000 • Fax: (802) 457-4004 • www.jewishlights.com
Credit card orders: **(800) 962-4544** (8:30AM–5:30PM EST Monday–Friday)
Generous discounts on quantity orders. SATISFACTION GUARANTEED. Prices subject to change.

Spiritual Practice—SkyLight Paths

Fly Fishing—The Sacred Art: Casting a Fly as a Spiritual Practice
by Rabbi Eric Eisenkramer and Rev. Michael Attas, MD
Illuminates what fly fishing can teach you about reflection, awe and wonder; the benefits of solitude; the blessing of community and the search for the Divine.
5½ x 8½, 192 pp (est), Quality PB, 978-1-59473-299-7 **$16.99**

Lectio Divina—The Sacred Art: Transforming Words & Images into Heart-Centered Prayer *by Christine Valters Paintner, PhD*
Expands the practice of sacred reading beyond scriptural texts and makes it accessible in contemporary life. 5½ x 8½, 240 pp, Quality PB, 978-1-59473-300-0 **$16.99**

Haiku—The Sacred Art: A Spiritual Practice in Three Lines
by Margaret D. McGee 5½ x 8½, 192 pp, Quality PB, 978-1-59473-269-0 **$16.99**

Dance—The Sacred Art: The Joy of Movement as a Spiritual Practice
by Cynthia Winton-Henry 5½ x 8½, 224 pp, Quality PB, 978-1-59473-268-3 **$16.99**

Spiritual Adventures in the Snow: Skiing & Snowboarding as Renewal for Your Soul
by Dr. Marcia McFee and Rev. Karen Foster; Foreword by Paul Arthur
5½ x 8½, 208 pp, Quality PB, 978-1-59473-270-6 **$16.99**

Divining the Body: Reclaim the Holiness of Your Physical Self *by Jan Phillips*
8 x 8, 256 pp, Quality PB, 978-1-59473-080-1 **$16.99**

Everyday Herbs in Spiritual Life: A Guide to Many Practices
by Michael J. Caduto; Foreword by Rosemary Gladstar
7 x 9, 208 pp, 20+ b/w illus., Quality PB, 978-1-59473-174-7 **$16.99**

Giving—The Sacred Art: Creating a Lifestyle of Generosity
by Lauren Tyler Wright 5½ x 8½, 208 pp, Quality PB, 978-1-59473-224-9 **$16.99**

Hospitality—The Sacred Art: Discovering the Hidden Spiritual Power of Invitation and Welcome *by Rev. Nanette Sawyer; Foreword by Rev. Dirk Ficca*
5½ x 8½, 208 pp, Quality PB, 978-1-59473-228-7 **$16.99**

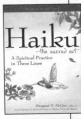

Labyrinths from the Outside In: Walking to Spiritual Insight—A Beginner's Guide
by Donna Schaper and Carole Ann Camp
6 x 9, 208 pp, b/w illus. and photos, Quality PB, 978-1-893361-18-8 **$16.95**

Practicing the Sacred Art of Listening: A Guide to Enrich Your Relationships and Kindle Your Spiritual Life *by Kay Lindahl* 8 x 8, 176 pp, Quality PB, 978-1-893361-85-0 **$16.95**

Recovery—The Sacred Art: The Twelve Steps as Spiritual Practice *by Rami Shapiro; Foreword by Joan Borysenko, PhD* 5½ x 8½, 240 pp, Quality PB, 978-1-59473-259-1 **$16.99**

Running—The Sacred Art: Preparing to Practice *by Dr. Warren A. Kay; Foreword by Kristin Armstrong* 5½ x 8½, 160 pp, Quality PB, 978-1-59473-227-0 **$16.99**

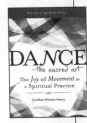

The Sacred Art of Chant: Preparing to Practice
by Ana Hernández 5½ x 8½, 192 pp, Quality PB, 978-1-59473-036-8 **$15.99**

The Sacred Art of Fasting: Preparing to Practice
by Thomas Ryan, CSP 5½ x 8½, 192 pp, Quality PB, 978-1-59473-078-8 **$15.99**

The Sacred Art of Forgiveness: Forgiving Ourselves and Others through God's Grace
by Marcia Ford 8 x 8, 176 pp, Quality PB, 978-1-59473-175-4 **$18.99**

The Sacred Art of Listening: Forty Reflections for Cultivating a Spiritual Practice
by Kay Lindahl; Illus. by Amy Schnapper 8 x 8, 160 pp, b/w illus., Quality PB, 978-1-893361-44-7 **$16.99**

The Sacred Art of Lovingkindness: Preparing to Practice
by Rabbi Rami Shapiro; Foreword by Marcia Ford 5½ x 8½, 176 pp, Quality PB, 978-1-59473-151-8 **$16.99**

Sacred Attention: A Spiritual Practice for Finding God in the Moment
by Margaret D. McGee 6 x 9, 144 pp, Quality PB, 978-1-59473-291-1 **$16.99**

Soul Fire: Accessing Your Creativity
by Thomas Ryan, CSP 6 x 9, 160 pp, Quality PB, 978-1-59473-243-0 **$16.99**

Thanking & Blessing—The Sacred Art: Spiritual Vitality through Gratefulness
by Jay Marshall, PhD; Foreword by Philip Gulley 5½ x 8½, 176 pp, Quality PB, 978-1-59473-231-7 **$16.99**

Judaism / Christianity / Islam / Interfaith

Christians & Jews—Faith to Faith: Tragic History, Promising Present, Fragile Future *by Rabbi James Rudin*
A probing examination of Christian-Jewish relations that looks at the major issues facing both faith communities. 6 x 9, 288 pp, HC, 978-1-58023-432-0 **$24.99***

Getting to the Heart of Interfaith
The Eye-Opening, Hope-Filled Friendship of a Pastor, a Rabbi & an Imam
by Pastor Don Mackenzie, Rabbi Ted Falcon and Imam Jamal Rahman
Offers many insights and encouragements for individuals and groups who want to tap into the promise of interfaith dialogue. 6 x 9, 192 pp, Quality PB, 978-1-59473-263-8 **$16.99**

Hearing the Call across Traditions: Readings on Faith and Service
Edited by Adam Davis; Foreword by Eboo Patel
Explores the connections between faith, service and social justice through the prose, verse and sacred texts of the world's great faith traditions.
6 x 9, 352 pp, Quality PB, 978-1-59473-303-1 **$18.99**; HC, 978-1-59473-264-5 **$29.99**

How to Do Good & Avoid Evil: A Global Ethic from the Sources of Judaism
by Hans Küng and Rabbi Walter Homolka; Translated by Rev. Dr. John Bowden
6 x 9, 224 pp, HC, 978-1-59473-255-3 **$19.99**

Blessed Relief: What Christians Can Learn from Buddhists about Suffering
by Gordon Peerman 6 x 9, 208 pp, Quality PB, 978-1-59473-252-2 **$16.99**

The Changing Christian World: A Brief Introduction for Jews
by Rabbi Leonard A. Schoolman 5½ x 8½, 176 pp, Quality PB, 978-1-58023-344-6 **$16.99***

Christians & Jews in Dialogue: Learning in the Presence of the Other *by Mary C. Boys and Sara S. Lee; Foreword by Dorothy C. Bass* 6 x 9, 240 pp, Quality PB, 978-1-59473-254-6 **$18.99**

Disaster Spiritual Care: Practical Clergy Responses to Community, Regional and National Tragedy *Edited by Rabbi Stephen B. Roberts, BCJC, and Rev. Willard W.C. Ashley, Sr., DMin, DH*
6 x 9, 384 pp, HC, 978-1-59473-240-9 **$50.00**

InterActive Faith: The Essential Interreligious Community-Building Handbook
Edited by Rev. Bud Heckman with Rori Picker Neiss; Foreword by Rev. Dirk Ficca
6 x 9, 304 pp, Quality PB, 978-1-59473-273-7 **$16.99**; HC, 978-1-59473-237-9 **$29.99**

The Jewish Approach to God: A Brief Introduction for Christians
by Rabbi Neil Gillman, PhD 5½ x 8½, 192 pp, Quality PB, 978-1-58023-190-9 **$16.95***

The Jewish Approach to Repairing the World (Tikkun Olam): A Brief Introduction for Christians *by Rabbi Elliot N. Dorff, PhD, with Rev. Cory Willson*
5½ x 8½, 256 pp, Quality PB, 978-1-58023-349-1 **$16.99***

The Jewish Connection to Israel, the Promised Land: A Brief Introduction for Christians *by Rabbi Eugene Korn, PhD* 5½ x 8½, 192 pp, Quality PB, 978-1-58023-318-7 **$14.99***

Jewish Holidays: A Brief Introduction for Christians *by Rabbi Kerry M. Olitzky and Rabbi Daniel Judson* 5½ x 8½, 176 pp, Quality PB, 978-1-58023-302-6 **$16.99***

Jewish Ritual: A Brief Introduction for Christians
by Rabbi Kerry M. Olitzky and Rabbi Daniel Judson 5½ x 8½, 144 pp, Quality PB, 978-1-58023-210-4 **$14.99***

Jewish Spirituality: A Brief Introduction for Christians *by Rabbi Lawrence Kushner*
5½ x 8½, 112 pp, Quality PB, 978-1-58023-150-3 **$12.95***

A Jewish Understanding of the New Testament *by Rabbi Samuel Sandmel;*
New preface by Rabbi David Sandmel 5½ x 8½, 368 pp, Quality PB, 978-1-59473-048-1 **$19.99***

Modern Jews Engage the New Testament: Enhancing Jewish Well-Being in a Christian Environment *by Rabbi Michael J. Cook, PhD* 6 x 9, 416 pp, HC, 978-1-58023-313-2 **$29.99***

Talking about God: Exploring the Meaning of Religious Life with Kierkegaard, Buber, Tillich and Heschel *by Daniel F. Polish, PhD* 6 x 9, 160 pp, Quality PB, 978-1-59473-272-0 **$16.99**

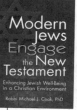

We Jews and Jesus: Exploring Theological Differences for Mutual Understanding
by Rabbi Samuel Sandmel; New preface by Rabbi David Sandmel
6 x 9, 192 pp, Quality PB, 978-1-59473-208-9 **$16.99**

Who Are the *Real* Chosen People? The Meaning of Chosenness in Judaism, Christianity and Islam *by Reuven Firestone, PhD*
6 x 9, 176 pp, Quality PB, 978-1-59473-290-4 **$16.99**; HC, 978-1-59473-248-5 **$21.99**

* A book from Jewish Lights, SkyLight Paths' sister imprint

About SKYLIGHT PATHS Publishing

SkyLight Paths Publishing is creating a place where people of different spiritual traditions come together for challenge and inspiration, a place where we can help each other understand the mystery that lies at the heart of our existence.

Through spirituality, our religious beliefs are increasingly becoming a part of our lives—rather than *apart* from our lives. While many of us may be more interested than ever in spiritual growth, we may be less firmly planted in traditional religion. Yet, we do want to deepen our relationship to the sacred, to learn from our own as well as from other faith traditions, and to practice in new ways.

SkyLight Paths sees both believers and seekers as a community that increasingly transcends traditional boundaries of religion and denomination—people wanting to learn from each other, *walking together, finding the way.*

For your information and convenience, at the back of this book we have provided a list of other SkyLight Paths books you might find interesting and useful. They cover the following subjects:

Buddhism / Zen	Global Spiritual	Monasticism
Catholicism	Perspectives	Mysticism
Children's Books	Gnosticism	Poetry
Christianity	Hinduism /	Prayer
Comparative	Vedanta	Religious Etiquette
Religion	Inspiration	Retirement
Current Events	Islam / Sufism	Spiritual Biography
Earth-Based	Judaism	Spiritual Direction
Spirituality	Kabbalah	Spirituality
Enneagram	Meditation	Women's Interest
	Midrash Fiction	Worship

Or phone, fax, mail or e-mail to: SKYLIGHT PATHS Publishing
Sunset Farm Offices, Route 4 • P.O. Box 237 • Woodstock, Vermont 05091
Tel: (802) 457-4000 • Fax: (802) 457-4004 • www.skylightpaths.com
Credit card orders: (800) 962-4544 (8:30AM–5:30PM EST Monday–Friday)
Generous discounts on quantity orders. SATISFACTION GUARANTEED. Prices subject to change.

About Jewish Lights

People of all faiths and backgrounds yearn for books that attract, engage, educate, and spiritually inspire.

Our principal goal is to stimulate thought and help all people learn about who the Jewish People are, where they come from, and what the future can be made to hold. While people of our diverse Jewish heritage are the primary audience, our books speak to people in the Christian world as well and will broaden their understanding of Judaism and the roots of their own faith.

We bring to you authors who are at the forefront of spiritual thought and experience. While each has something different to say, they all say it in a voice that you can hear.

Our books are designed to welcome you and then to engage, stimulate, and inspire. We judge our success not only by whether or not our books are beautiful and commercially successful, but by whether or not they make a difference in your life.

For your information and convenience, at the back of this book we have provided a list of other Jewish Lights books you might find interesting and useful. They cover all the categories of your life:

Bar/Bat Mitzvah	Life Cycle
Bible Study / Midrash	Meditation
Children's Books	Men's Interest
Congregation Resources	Parenting
Current Events / History	Prayer / Ritual / Sacred Practice
Ecology / Environment	Social Justice
Fiction: Mystery, Science Fiction	Spirituality
Grief / Healing	Theology / Philosophy
Holidays / Holy Days	Travel
Inspiration	Twelve Steps
Kabbalah / Mysticism / Enneagram	Women's Interest

Stuart M. Matlins, Publisher

Or phone, fax, mail or e-mail to: **JEWISH LIGHTS Publishing**
Sunset Farm Offices, Route 4 • P.O. Box 237 • Woodstock, Vermont 05091
Tel: (802) 457-4000 • Fax: (802) 457-4004 • www.jewishlights.com
Credit card orders: (800) 962-4544 (8:30AM–5:30PM EST Monday–Friday)
Generous discounts on quantity orders. SATISFACTION GUARANTEED. Prices subject to change.

For more information about each book, visit our website at www.jewishlights.com